P9-CKJ-878

THE MUSCULOSKELETAL SYSTEM

THE ENCYCLOPEDIA OF
HEALTH

THE HEALTHY BODY

Dale C. Garell, M.D. • General Editor

THE MUSCULOSKELETAL SYSTEM

Brian Feinberg

Introduction by C. Everett Koop, M.D., Sc.D.

former Surgeon General, U. S. Public Health Service

CHELSEA HOUSE PUBLISHERS

New York • Philadelphia

...H is to provide general information in ...sychology, and related medical issues. ...l to take the place of the professional advice of a physician or other health care professional.

ON THE COVER: Illustration of the back muscles from *Gray's Anatomy*.

CHELSEA HOUSE PUBLISHERS
EDITORIAL DIRECTOR Richard Rennert
EXECUTIVE MANAGING EDITOR Karyn Gullen Browne
EXECUTIVE EDITOR Sean Dolan
COPY CHIEF Philip Koslow
PICTURE EDITOR Adrian G. Allen
ART DIRECTOR Nora Wertz
MANUFACTURING DIRECTOR Gerald Levine
SYSTEMS MANAGER Lindsey Ottman
PRODUCTION COORDINATOR Marie Claire Cebrián-Ume

The Encyclopedia of Health
SENIOR EDITOR Kenneth W. Lane

Staff for THE MUSCULOSKELETAL SYSTEM
COPY EDITOR Danielle Janusz
EDITORIAL ASSISTANT Laura Petermann
PICTURE RESEARCHER Sandy Jones
DESIGNER Robert Yaffe

First Printing
1 3 5 7 9 8 6 4 2

Library of Congress Cataloging-in-Publication Data

Feinberg, Brian
The Musculoskeletal System/by Brian Feinberg; introduction by C. Everett Koop.
p. cm.—(The Encyclopedia of health)
Includes bibliographical references and index.
Summary: An examination of the musculoskeletal system, including its structure, functions, and disorders.
ISBN 0-7910-0028-1
0-7910-0463-5 (pbk.)
1. Musculoskeletal system—Physiology—Juvenile literature. 2. Musculoskeletal system—Diseases—Juvenile literature. [1. Muscular system. 2. Skeleton.] I. Title. II. Series. 92-21956
QP301.F425 1993 CIP
612.7'4—dc20 AC

CONTENTS

"Prevention and Education:
The Keys to Good Health"—
C. Everett Koop, M.D., Sc.D. 7

Foreword—Dale C. Garell, M.D. 11

1 The History of Musculoskeletal Study 13

2 The Skeletal System 23

3 Bone Growth and Development 35

4 The Muscular System 43

5 Skeletal Muscle and Exercise 59

6 Skeletal Injuries 65

7 Skeletal and Joint Disorders 73

8 Muscular Disorders and Injuries 87

Appendix: For More Information 96

Further Reading 98

Glossary 100

Index 107

THE ENCYCLOPEDIA OF
HEALTH

THE HEALTHY BODY

The Circulatory System
Dental Health
The Digestive System
The Endocrine System
Exercise
Genetics & Heredity
The Human Body: An Overview
Hygiene
The Immune System
Memory & Learning
The Musculoskeletal System
The Nervous System
Nutrition
The Reproductive System
The Respiratory System
The Senses
Sleep
Speech & Hearing
Sports Medicine
Vision
Vitamins & Minerals

THE LIFE CYCLE

Adolescence
Adulthood
Aging
Childhood
Death & Dying
The Family
Friendship & Love
Pregnancy & Birth

MEDICAL ISSUES

Careers in Health Care
Environmental Health
Folk Medicine
Health Care Delivery
Holistic Medicine
Medical Ethics
Medical Fakes & Frauds
Medical Technology
Medicine & the Law
Occupational Health
Public Health

PSYCHOLOGICAL DISORDERS AND THEIR TREATMENT

Anxiety & Phobias
Child Abuse
Compulsive Behavior
Delinquency & Criminal Behavior
Depression
Diagnosing & Treating Mental Illness
Eating Habits & Disorders
Learning Disabilities
Mental Retardation
Personality Disorders
Schizophrenia
Stress Management
Suicide

MEDICAL DISORDERS AND THEIR TREATMENT

AIDS
Allergies
Alzheimer's Disease
Arthritis
Birth Defects
Cancer
The Common Cold
Diabetes
Emergency Medicine
Gynecological Disorders
Headaches
The Hospital
Kidney Disorders
Medical Diagnosis
The Mind-Body Connection
Mononucleosis and Other Infectious Diseases
Nuclear Medicine
Organ Transplants
Pain
Physical Handicaps
Poisons & Toxins
Prescription & OTC Drugs
Sexually Transmitted Diseases
Skin Disorders
Stroke & Heart Disease
Substance Abuse
Tropical Medicine

PREVENTION AND EDUCATION: THE KEYS TO GOOD HEALTH

C. Everett Koop, M.D., Sc.D.
former Surgeon General,
U.S. Public Health Service

The issue of health education has received particular attention in recent years because of the presence of AIDS in the news. But our response to this particular tragedy points up a number of broader issues that doctors, public health officials, educators, and the public face. In particular, it points up the necessity for sound health education for citizens of all ages.

Over the past 25 years this country has been able to bring about dramatic declines in the death rates for heart disease, stroke, accidents, and for people under the age of 45, cancer. Today, Americans generally eat better and take better care of themselves than ever before. Thus, with the help of modern science and technology, they have a better chance of surviving serious—even catastrophic—illnesses. That's the good news.

But, like every phonograph record, there's a flip side, and one with special significance for young adults. According to a report issued in 1979 by Dr. Julius Richmond, my predecessor as Surgeon General, Americans aged 15 to 24 had a higher death rate in 1979 than they did 20 years earlier. The causes: violent death and injury, alcohol and drug abuse, unwanted pregnancies, and sexually transmitted diseases. Adolescents are particularly vulnerable because they are beginning to explore their own sexuality and perhaps to experiment with drugs. The need for educating young people is critical, and the price of neglect is high.

Yet even for the population as a whole, our health is still far from what it could be. Why? A 1974 Canadian government report attributed all death and disease to four broad elements: inadequacies in the health care system, behavioral factors or unhealthy life-styles, environmental hazards, and human biological factors.

To be sure, there are diseases that are still beyond the control of even our advanced medical knowledge and techniques. And despite yearnings that are as old as the human race itself, there is no "fountain of youth" to ward off aging and death. Still, there is a solution to many of the problems that undermine sound health. In a word, that solution is prevention. Prevention, which includes health promotion and education, saves lives, improves the quality of life, and in the long run, saves money.

In the United States, organized public health activities and preventive medicine have a long history. Important milestones in this country or foreign breakthroughs adopted in the United States include the improvement of sanitary procedures and the development of pasteurized milk in the late 19th century and the introduction in the mid-20th century of effective vaccines against polio, measles, German measles, mumps, and other once-rampant diseases. Internationally, organized public health efforts began on a wide-scale basis with the International Sanitary Conference of 1851, to which 12 nations sent representatives. The World Health Organization, founded in 1948, continues these efforts under the aegis of the United Nations, with particular emphasis on combating communicable diseases and the training of health care workers.

Despite these accomplishments, much remains to be done in the field of prevention. For too long, we have had a medical care system that is science- and technology-based, focused, essentially, on illness and mortality. It is now patently obvious that both the social and the economic costs of such a system are becoming insupportable.

Implementing prevention—and its corollaries, health education and promotion—is the job of several groups of people.

First, the medical and scientific professions need to continue basic scientific research, and here we are making considerable progress. But increased concern with prevention will also have a decided impact on how primary care doctors practice medicine. With a shift to health-based rather than morbidity-based medicine, the role of the "new physician" will include a healthy dose of patient education.

Second, practitioners of the social and behavioral sciences—psychologists, economists, city planners—along with lawyers, business leaders, and government officials—must solve the practical and ethical dilemmas confronting us: poverty, crime, civil rights, literacy, education, employment, housing, sanitation, environmental protection, health care delivery systems, and so forth. All of these issues affect public health.

Third is the public at large. We'll consider that very important group in a moment.

Fourth, and the linchpin in this effort, is the public health profession—doctors, epidemiologists, teachers—who must harness the professional expertise of the first two groups and the common sense and cooperation of the third, the public. They must define the problems statistically and qualitatively and then help us set priorities for finding the solutions.

To a very large extent, improving those statistics is the responsibility of every individual. So let's consider more specifically what the role of the individual should be and why health education is so important to that role. First, and most obvious, individuals can protect themselves from illness and injury and thus minimize their need for professional medical care. They can eat nutritious food; get adequate exercise; avoid tobacco, alcohol, and drugs; and take prudent steps to avoid accidents. The proverbial "apple a day keeps the doctor away" is not so far from the truth, after all.

Second, individuals should actively participate in their own medical care. They should schedule regular medical and dental checkups. Should they develop an illness or injury, they should know when to treat themselves and when to seek professional help. To gain the maximum benefit from any medical treatment that they do require, individuals must become partners in that treatment. For instance, they should understand the effects and side effects of medications. I counsel young physicians that there is no such thing as too much information when talking with patients. But the corollary is the patient must know enough about the nuts and bolts of the healing process to understand what the doctor is telling him or her. That is at least partially the patient's responsibility.

Education is equally necessary for us to understand the ethical and public policy issues in health care today. Sometimes individuals will encounter these issues in making decisions about their own treatment or that of family members. Other citizens may encounter them as jurors in medical malpractice cases. But we all become involved, indirectly, when we elect our public officials, from school board members to the president. Should surrogate parenting be legal? To what extent is drug testing desirable, legal, or necessary? Should there be public funding for family planning, hospitals, various types of medical research, and other medical care for the indigent? How should we allocate scant technological resources, such as kidney dialysis and organ transplants? What is the proper role of government in protecting the rights of patients?

What are the broad goals of public health in the United States today? In 1980, the Public Health Service issued a report aptly entitled *Promoting Health—Preventing Disease: Objectives for the Nation*. This report expressed its goals in terms of mortality and in terms of intermediate goals in education and health improvement. It identified 15 major concerns: controlling high blood pressure; improving family planning; improving pregnancy care and infant health; increasing the rate of immunization; controlling sexually transmitted diseases; controlling the presence of toxic agents and radiation in the environment; improving occupational safety and health; preventing accidents; promoting water fluoridation and dental health; controlling infectious diseases; decreasing smoking; decreasing alcohol and drug abuse; improving nutrition; promoting physical fitness and exercise; and controlling stress and violent behavior.

For healthy adolescents and young adults (ages 15 to 24), the specific goal was a 20% reduction in deaths, with a special focus on motor vehicle injuries and alcohol and drug abuse. For adults (ages 25 to 64), the aim was 25% fewer deaths, with a concentration on heart attacks, strokes, and cancers.

Smoking is perhaps the best example of how individual behavior can have a direct impact on health. Today, cigarette smoking is recognized as the single most important preventable cause of death in our society. It is responsible for more cancers and more cancer deaths than any other known agent; is a prime risk factor for heart and blood vessel disease, chronic bronchitis, and emphysema; and is a frequent cause of complications in pregnancies and of babies born prematurely, underweight, or with potentially fatal respiratory and cardiovascular problems.

Since the release of the Surgeon General's first report on smoking in 1964, the proportion of adult smokers has declined substantially, from 43% in 1965 to 30.5% in 1985. Since 1965, 37 million people have quit smoking. Although there is still much work to be done if we are to become a "smoke-free society," it is heartening to note that public health and public education efforts—such as warnings on cigarette packages and bans on broadcast advertising—have already had significant effects.

In 1835, Alexis de Tocqueville, a French visitor to America, wrote, "In America the passion for physical well-being is general." Today, as then, health and fitness are front-page items. But with the greater scientific and technological resources now available to us, we are in a far stronger position to make good health care available to everyone. And with the greater technological threats to us as we approach the 21st century, the need to do so is more urgent than ever before. Comprehensive information about basic biology, preventive medicine, medical and surgical treatments, and related ethical and public policy issues can help you arm yourself with the knowledge you need to be healthy throughout your life.

FOREWORD

Dale C. Garell, M.D.

Advances in our understanding of health and disease during the 20th century have been truly remarkable. Indeed, it could be argued that modern health care is one of the greatest accomplishments in all of human history. In the early 20th century, improvements in sanitation, water treatment, and sewage disposal reduced death rates and increased longevity. Previously untreatable illnesses can now be managed with antibiotics, immunizations, and modern surgical techniques. Discoveries in the fields of immunology, genetic diagnosis, and organ transplantation are revolutionizing the prevention and treatment of disease. Modern medicine is even making inroads against cancer and heart disease, two of the leading causes of death in the United States.

Although there is much to be proud of, medicine continues to face enormous challenges. Science has vanquished diseases such as smallpox and polio, but new killers, most notably AIDS, confront us. Moreover, we now victimize ourselves with what some have called "diseases of choice," or those brought on by drug and alcohol abuse, bad eating habits, and mismanagement of the stresses and strains of contemporary life. The very technology that is doing so much to prolong life has brought with it previously unimaginable ethical dilemmas related to issues of death and dying. The rising cost of health care is a matter of central concern to us all. And violence in the form of automobile accidents, homicide, and suicide remains the major killer of young adults.

In the past, most people were content to leave health care and medical treatment in the hands of professionals. But since the 1960s, the consumer of medical care—that is, the patient—has assumed an increasingly central role in the management of his or her own health. There has also been a new emphasis placed on prevention: People are recognizing that their own actions can help prevent many of the conditions that have caused death and disease in the past. This accounts for the growing commitment to good nutrition and regular exercise, for the increasing number of people who are choosing not to smoke, and for a new moderation in people's drinking habits.

People want to know more about themselves and their own health. They are curious about their body: its anatomy, physiology, and biochemistry. They want to keep up with rapidly evolving medical technologies and procedures. They are willing to educate themselves about common disorders and diseases so that they can be full partners in their own health care.

THE ENCYCLOPEDIA OF HEALTH is designed to provide the basic knowledge that readers will need if they are to take significant responsibility for their own health. It is also meant to serve as a frame of reference for further study and exploration. The encyclopedia is divided into five subsections: The Healthy Body; The Life Cycle; Medical Disorders & Their Treatment; Psychological Disorders & Their Treatment; and Medical Issues. For each topic covered by the encyclopedia, we present the essential facts about the relevant biology; the symptoms, diagnosis, and treatment of common diseases and disorders; and ways in which you can prevent or reduce the severity of health problems when that is possible. The encyclopedia also projects what may lie ahead in the way of future treatment or prevention strategies.

The broad range of topics and issues covered in the encyclopedia reflects that human health encompasses physical, psychological, social, environmental, and spiritual well-being. Just as the mind and the body are inextricably linked, so, too, is the individual an integral part of the wider world that comprises his or her family, society, and environment. To discuss health in its broadest aspect it is necessary to explore the many ways in which it is connected to such fields as law, social science, public policy, economics, and even religion. And so, the encyclopedia is meant to be a bridge between science, medical technology, the world at large, and you. I hope that it will inspire you to pursue in greater depth particular areas of interest and that you will take advantage of the suggestions for further reading and the lists of resources and organizations that can provide additional information.

CHAPTER 1

THE HISTORY OF MUSCULOSKELETAL STUDY

The Greek scientist Galen. His observations, based largely on the dissection of apes, led him to conclude that movement of the body and its various parts results from the contraction of muscles.

The *musculoskeletal system*, the body's interdependent network of muscles and bones, provides the body with shape, support, and the machinery for movement, in addition to playing an important role in the chemistry of life. Although these functions may seem obvious to today's anatomy student, medicine's insight into this essential system has taken thousands of years to develop.

ANCIENT INSIGHTS

Despite the excellent re-creation of human musculature found in ancient Greek sculpture, the creators of these remarkable works understood little about the function of muscle. The early Greeks did not believe that muscles played a role in movement, but instead considered them merely supporting material or stuffing between the skin and bones. The Greeks were convinced that it was the *tendons* (tough cords of fiber connecting muscle to bone) that caused the joints to bend.

Historians believe that the first clear description of muscle structure did not appear until the 2nd century A.D., when the Greek physician Rufus of Ephesus began dissecting and studying the internal structure of apes. He discovered that the muscles contain tendons, veins, and arteries, and also realized, perhaps for the first time in history, that muscles are used in voluntary movement. (Voluntary movement refers to motions—such as lifting the arms or wiggling the toes—that an individual performs at will.) Rufus apparently made another insight by realizing that nerves play a role in muscle movement.

Although he also described the bones, Rufus did not completely understand their anatomy, calling them bloodless and insensitive when in fact they are laced with nerves and blood vessels. But he clearly realized that the skeleton serves as a supporting framework for the body.

Muscle theory was advanced even further during the 2nd century by the Greek scientist Galen (A.D. 129–ca. 199), whose pioneering work influenced science for a millennium and a half. By pulling various tendons, he learned which muscles control various movements. Like Rufus, Galen based his work mainly on the examination of apes. His thorough dissections allowed him to describe more than 300 distinct muscles, and his research led him to conclude that movement results from the muscles' ability to contract. But not even Galen's conclusions were completely accurate. For example, his descriptions of where individual muscles begin and end were not totally correct, and he did not consider the heart to be a muscle (although he did classify it as musclelike).

A student of the skeleton as well as the muscles, Galen mistakenly believed that the *marrow* (the soft tissue within bones) was a source of nutrition for bone, a theory suggested earlier by the Greek physician Hippocrates (ca. 460–ca. 377 B.C.).

THE MIDDLE AGES AND THE RENAISSANCE

Anatomical Art

It was not until the 15th century that the first truly accurate and detailed drawings of bones were produced. By this time artists saw the need to perfect their skills through an understanding of human anatomy. Not surprisingly, the individual who best accomplished this integration of art and science was Leonardo da Vinci (1452–1519). An artist who was also an architect, engineer, and scientist, da Vinci became familiar with the body's internal structure by performing dissections and producing almost 1,000 chalk drawings and sketches of the human anatomy. He

A study of the bone structure of the leg by the Renaissance scientist and artist Leonardo da Vinci. Da Vinci made more than 1,000 such sketches and chalk drawings of dissections of the body.

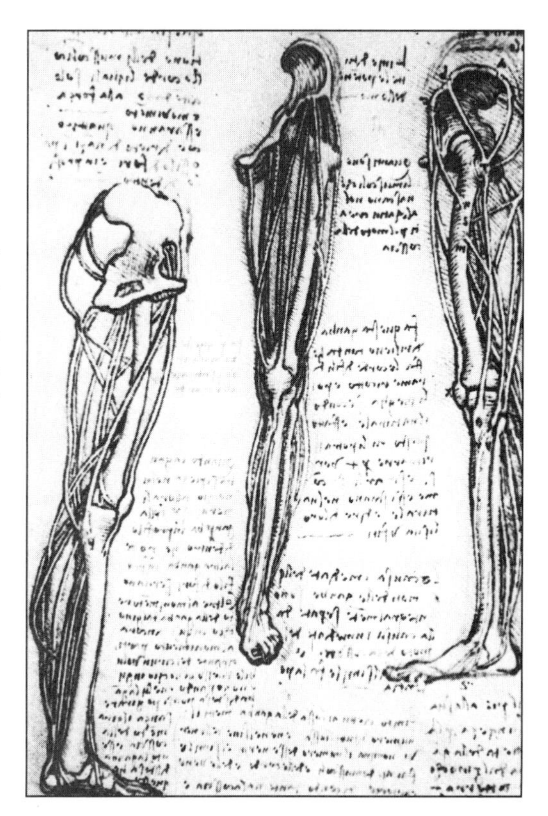

not only diagrammed the intricate joint-by-joint connections between bones but also attempted to reveal the arrangement of the muscles and veins within the body. Beyond this, he examined the function of the lungs, heart, and nervous system and was the first artist to draw anatomy in cross section. His effort, however, was not immediately recognized, because da Vinci's drawings were lost for almost 200 years. They were rediscovered around 1700 by the Englishman William Hunter, allowing anatomy students to at last appreciate da Vinci's accomplishments.

Another Renaissance artist, Michelangelo (1475–1564), seems to have had a more immediate effect on anatomical science. His legendary painting entitled *The Last Judgement*, which decorates Rome's Sistine Chapel, depicts the martyred Saint Bartholomew holding a knife in one hand and his flayed skin in the other. The work was unveiled in 1541 and was followed, according to an article in the October 1954 issue of the *Journal of the History of Medicine*, by two centuries of imitation by anatomical artists. What was apparently the first such Michelangelo-inspired work came from the Spanish artist Gasparo Becerra, who

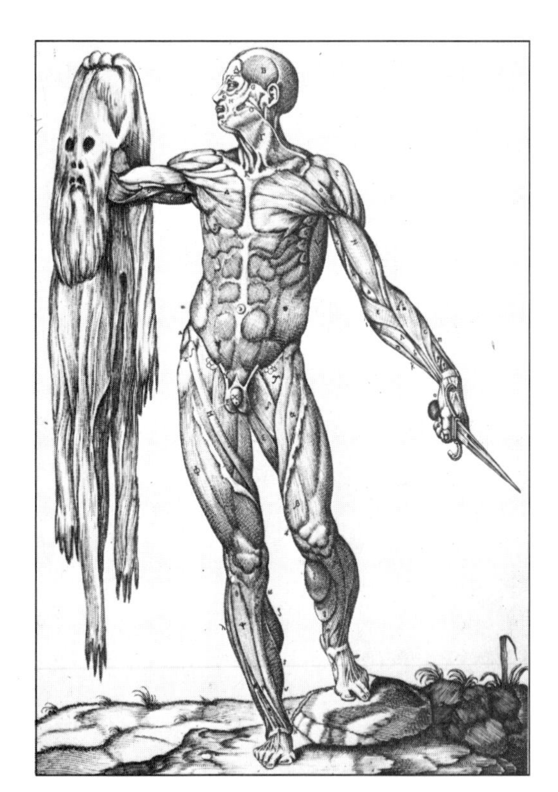

A drawing by the Spanish artist Gasparo Becerra of the muscles and connective tissues immediately beneath the skin, which, flayed and removed from the body, hangs limply on the left.

produced a detailed drawing of the human musculature for an anatomy book entitled *Historia de la composición del cuerpo humano*, published in Rome in 1556. In the drawing, a skinless man, whose muscles are well delineated, poses much like Michelangelo's Saint Bartholomew, a dagger in one hand and his flesh clasped firmly in the other. Similar sketches appeared in anatomical texts through 1728, when the Spanish work *Anatomía completa del hombre*, illustrated by the monk Mateo Antonio Irala, was published.

Bonesetters

Working in medieval and early modern Europe were numerous bonesetters. These primitive health care providers often practiced another trade—such as carpentry or blacksmithing—within their community, but earned extra income by setting broken bones. Many retired sailors also filled the role of bonesetter because such skills were required among a ship's crew.

A bonesetter's job was to treat fractures, sprains, and dislocated joints. Some bonesetters also tried to help patients suffering from various deformities by applying splints, braces, or stiff bandages to the deformed limbs. Eventually, as medicine advanced and physicians became skilled in dealing with skeletal injury and malformation, they assumed the bonesetter's function.

Andreas Vesalius

In 1543, the science of anatomy witnessed a great advance with the publication of the seven-volume *De Humani Corporis Fabrica* (The Fabric of the Human Body), by the Belgian anatomist Andreas Vesalius (1514–64). In that same year Vesalius also published the *Epitome*, an anatomy text for laymen. Both books contained detailed and—for the most part—accurate drawings by the Flemish artist Jan Stefan. The artwork included one of the best-known images in anatomy—that of a human skeleton leaning against a tomb as it contemplates a skull. The drawings also showed a gradual dissection of the musculature, revealing separate layers of muscle within the body.

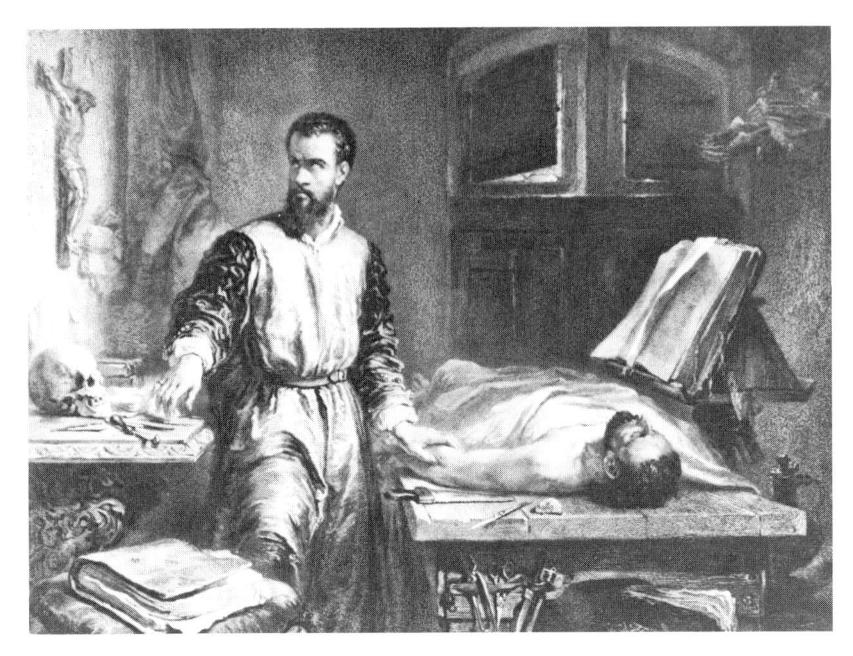

The Belgian anatomist Andreas Vesalius, shown here beginning a dissection, made great contributions to the modern scientific understanding of the body's internal structure.

Vesalius's quest to understand human anatomy resulted in a dramatic and very curious life. The 1929 book *The Struggle for Health*, by Richard H. Hoffmann, details the rise and fall of this talented scientist. As a young student Vesalius performed his first dissections on rodents, cats, and dogs—the only subjects then at his disposal—before his scientific talents led him first to Belgium's University of Louvain and then to further schooling in Paris. In France, Vesalius's advanced ideas brought him into conflict with a strict teacher named Jacobus Sylvius, who was loyal to the teachings of Galen. In 1536, because of the outbreak of war between France and Germany, Vesalius is reported to have returned to Louvain. Here, says Hoffmann's account, fate provided Vesalius with a great, if somewhat ghoulish, opportunity.

At a nearby execution ground, Vesalius and a friend happened upon the body of a notorious thief who had been burned to death and whose bones had been picked clean by scavenging birds. Realizing his opportunity, Vesalius stole what bones he could from the corpse and then returned on successive nights, each time removing a bit more of the skeleton and bringing it home until he had reassembled the unfortunate thief's skeleton in his room. When his impropriety was discovered, the young student was banished from Louvain. But by that time, he had gained a sound understanding of human skeletal structure.

Later, as a military surgeon, Vesalius further improved his anatomical knowledge and eventually used his advanced insight into the human body to become a renowned teacher. When the powerful duke of Florence became his patron, Vesalius was able to use the nobleman's influence to secure as many human bodies as he needed for study. In fact, while living in Venice, where the duke's influence was strong, Vesalius was permitted to suggest the method of death for condemned criminals, since certain forms of execution better served the anatomist when he later examined the bodies. When the duke died, however, Vesalius had to turn to grave robbing to meet his needs.

Ironically, Vesalius's greatest achievement also injured him professionally. In *De Humani Corporis Fabrica*, he was extremely critical of Galen, whose work at that time was still considered standard knowledge. As a result, Vesalius's own credibility was questioned. He was even attacked by his contentious former teacher Sylvius, and in frustration the anatomist gave up much of his research. Eventually, he became physician to King Philip II of Spain and later made a religious pilgrimage to the Middle East. (One story, perhaps false, claims that Vesalius was forced to make this pilgrimage in atonement for having dissected the body of a Spanish nobleman who, Vesalius discovered upon opening the subject, was still alive.) On his way home from the Middle East, Vesalius's ship was wrecked on the Greek island of Zante, where he died.

EARLY MODERN AND MODERN ADVANCES

The word *orthopedics*, which refers to the branch of surgery treating the bones and joints, was coined in 1741 by Nicholas André. The

Parisian physician, who wrote a book discussing the prevention and correction of musculoskeletal deformities in children, created the term from the Greek words *orthos* (straight) and *paidios* (child). Although he did not actually invent the science of orthopedics, André did help to advance it through his work.

Earlier in the 18th century, new insight was gained into the muscular system when the Dutch scientist Antonie van Leeuwênhoek (1632–1723) looked through a microscope to describe the bands that can be seen running vertically across *skeletal muscle* cells. The Italian scientist Giorgio Baglivi (1668–1707) became the first to note that skeletal muscle cells differ microscopically from the cells of *smooth* or involuntary muscle (muscle involved in movements the body performs automatically), which do not have a banded appearance. (Muscle cells will be discussed in detail in chapter 4.)

During the late 18th and early 19th centuries, the field of orthopedic surgery began to develop. Researchers, including British surgeon Percivall Pott (1714–88), made studies of spinal decay, recognizing the disorder that came to be known as *Pott's disease*—tuberculosis of the spine, caused by the bacteria *Mycobacterium tuberculosis*. During the first half of the 19th century, physicians developed a surgical treatment for *clubfoot*, a birth defect involving muscle and bone that often gives the foot a clublike appearance.

When quick-setting plaster of paris was introduced in the 1850s, physicians were better able to fashion the bandages used for making orthopedic corrections to the arms, legs, and other parts of the body. At about this time also, the British physician Hugh Owen Thomas (1834–91) developed the Thomas splint, an iron support for orthopedic treatment that is still used today. The study of bone was further advanced in 1868 when German scientist Ernst Neumann (1834–1918) first proposed a link between bone marrow and the formation of blood cells. (Bone marrow will be further discussed in chapter 2.)

By the end of the 19th century, orthopedic surgeons had become commonplace in the United States, and leading orthopedic centers existed in New York and Boston. Between 1911 and 1915, the American orthopedic surgeon Fred H. Albee (1876–1945) made im-

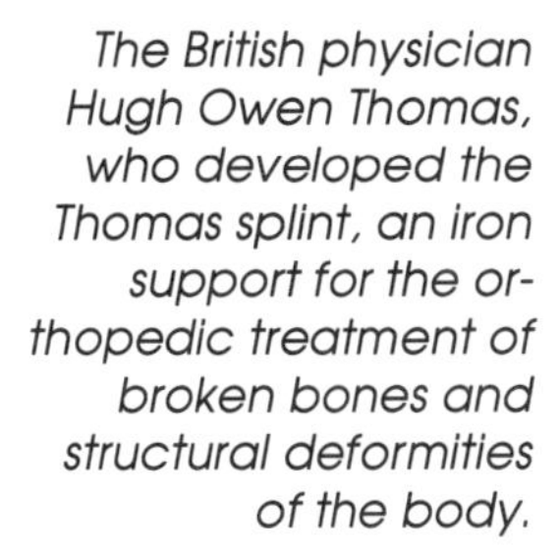

The British physician Hugh Owen Thomas, who developed the Thomas splint, an iron support for the orthopedic treatment of broken bones and structural deformities of the body.

portant advances in the use of bone grafts, in which bone is transplanted from another part of the patient to treat bone diseases, fractures, and deformities.

An important 19th-century advance in the study of muscles was made by German physiologist Hermann von Helmholtz (1821–94). Studying chemical reactions in frog muscles, he proposed in 1845 that such phenomena are the main source of body heat. In 1850 von Helmholtz also confirmed the theory, postulated a few years earlier, that muscle movement can lead to the formation of a type of acid—later identified as *lactic acid* (see chapter 5). By the late 19th century, the understanding of muscular diseases was improving as researchers first began to describe the different forms of *muscular dystrophy*, a hereditary disease that causes the progressive destruction of muscle tissue and its replacement by fat.

Orthopedics continued to expand after World War I as medical schools adopted programs in this field of medicine and specialists in orthopedics increasingly took over treatment that had previously been left to less specialized physicians.

Hermann von Helmholtz, a German physiologist, revealed that chemical reactions in muscles are the major source of heat generated by the body and confirmed that muscle movement leads to the formation of lactic acid.

The long history of musculoskeletal research has yielded invaluable knowledge not only of the overall anatomy of the human body but also of the cellular and chemical basis of muscles and bones and of the disorders that afflict them. These topics will be discussed in the ensuing chapters.

CHAPTER 2

THE SKELETAL SYSTEM

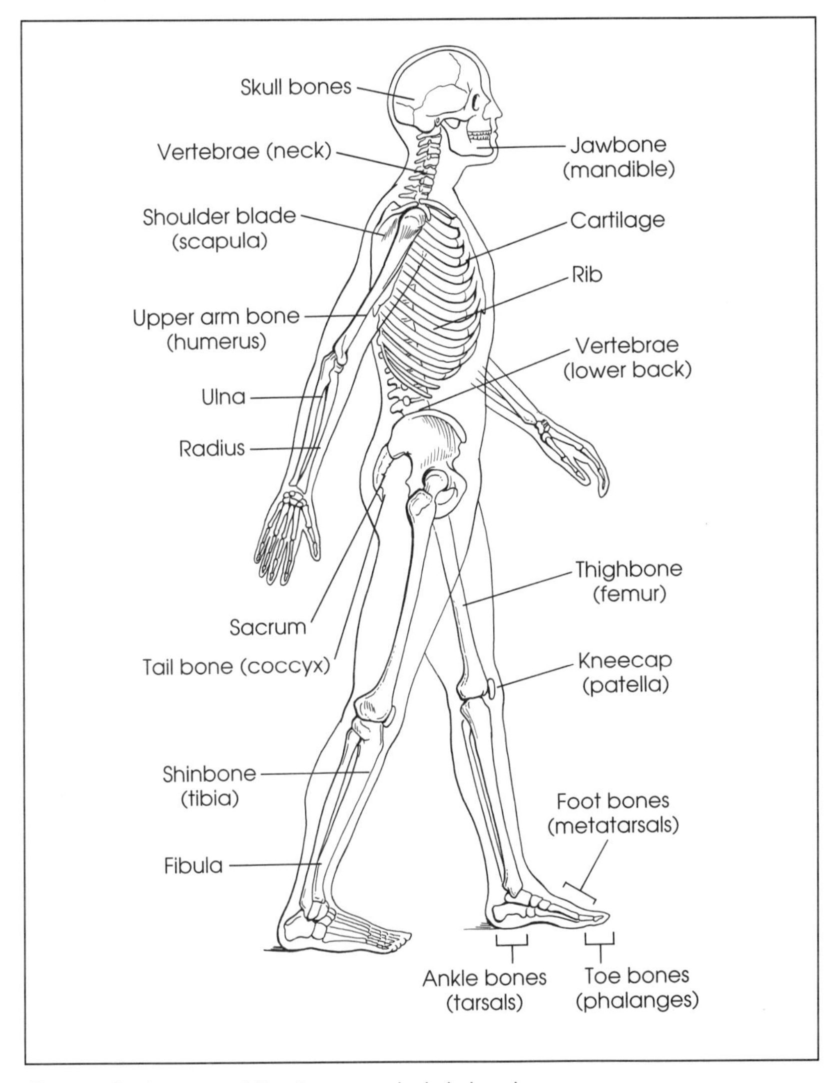

The major bones of the human skeletal system.

The adult human skeleton contains approximately 206 bones, although it may be surprising to learn that the number can vary from one individual to another. Besides providing the body with the rigid support that gives it its shape and permits it to carry its own weight, the skeleton also protects the body's internal organs. The bones are also essential to movement, working as levers that permit humans to do everything from raising their arms to walking across a room. Beyond this, the skeleton also plays a major role in the body's chemistry, or *metabolism*, by storing important nutrients. Additionally, by producing the blood cells that carry oxygen and nutrients through the body and fight infection, the skeleton serves yet another vital function.

SUPPORT, PROTECTION, AND MOVEMENT

Certain creatures, such as insects and shellfish, have developed *exoskeletons*, hard, protective coverings over the outside of their bodies. By contrast, *vertebrates*, the animals that have backbones, have their rigid supporting structures inside their bodies in the form of *endoskeletons*. Various parts of the endoskeleton shield different organs. The rib cage and shoulder girdle, for example, protect the heart and lungs; the skull safeguards the eyes, ears, and brain; and the bones of the spinal column, known as *vertebrae*, surround the spinal cord.

The same rigidity and hardness that give the skeleton its protective property also permit the bones—particularly those of the feet, legs, spine, and pelvis—to support the body's weight. As a result, humans can sit up or stand straight against the force of gravity that pulls downward on the body.

The rigidity of bone is also essential to movement. Every movement of the body is based on the principle of the lever. The bones act as the lever itself, and the joints act as the fulcrum on which the lever pivots. Muscle contraction provides the force on the lever. Depending on the nature of a specific movement, the bones and joints that participate in the movement may serve as first-, second-, or third-class levers.

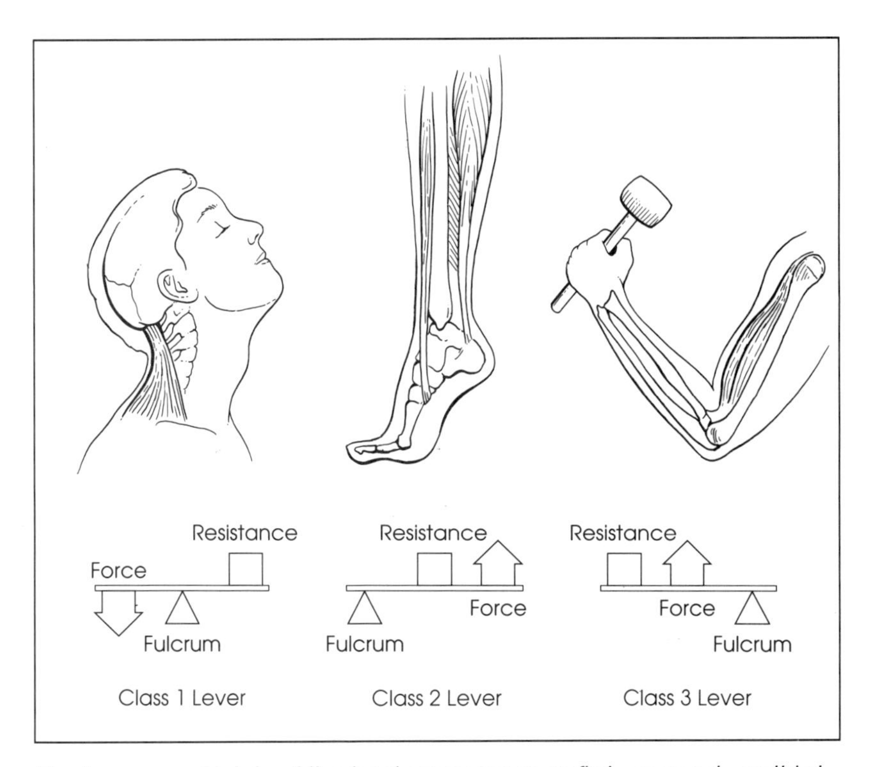

The bones and joints of the body may serve as first-, second-, or third-class levers. An example of a first-class lever is the upward pivoting of the head on the neck. The foot provides an example of a second-class lever, with the metatarsal bones in the ball of the foot serving as the fulcrum, or pivot point. The elbow, forearm, and hand represent a third-class lever. The class of lever depends upon the placement of the fulcrum, the resisting force, and the point where effort is applied.

STORAGE FUNCTION

As noted earlier, the bones play a vital role in storing nutrients for the body. One of these nutrients is the mineral calcium, which is necessary for the clotting of blood, the nerves' conduction of electrical impulses, and the contraction of muscles, and which is stored in the form of calcium salts. The two chief calcium salts in bone, calcium phosphate and calcium carbonate, are stored in the form of crystals of a substance

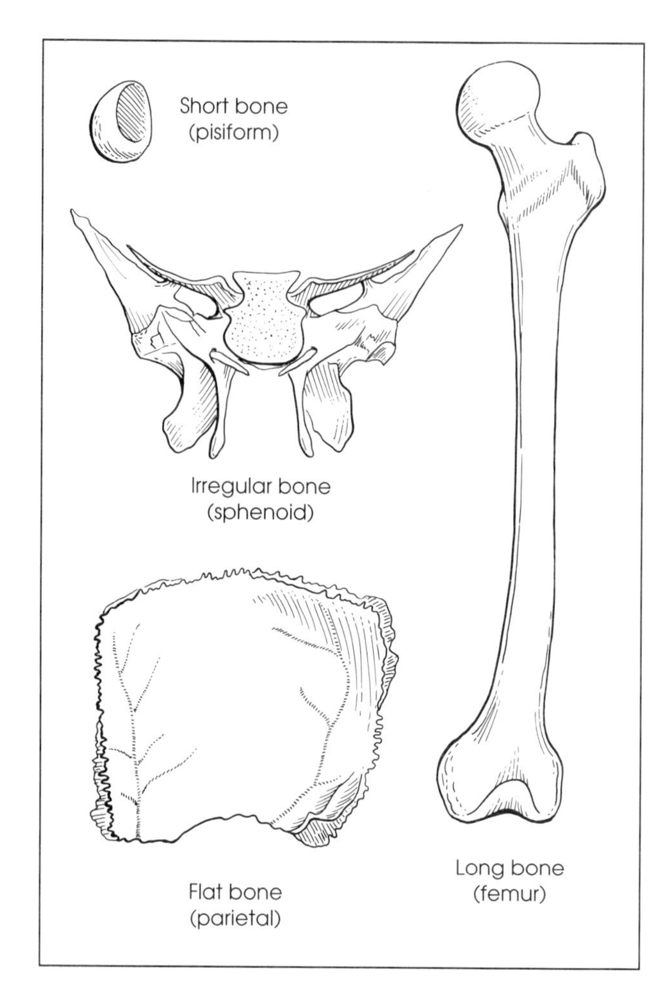

The shapes of bones differ and are matched to their functions. The femur is the long bone of the upper leg. The parietal bone of the skull provides a strong shield for the brain that lies beneath it. The sphenoid bone, located within the skull, supports the lower middle part of the brain and helps form the eye sockets. The pisiform bone is one of several bones located in the heel of the hand.

known as hydroxyapatite. This substance is part of the bone *matrix*—the material that fills the area between bone cells, much as mortar joins bricks. When the level of calcium in the blood falls below what the body needs, specialized bone cells known as *osteoclasts* break down bone matrix, releasing some of the skeleton's calcium supply, which then enters the blood to restore its calcium content. If the blood contains too much calcium, new bone tissue is formed to take up the excess in storage. Other minerals contained in the bone matrix include magnesium, potassium, and sodium, although not in as great quantities as calcium.

Fat is also stored within bone, in the tissue known as *yellow marrow*. This type of marrow is located in an inner area called the *medullary cavity*, which runs lengthwise through the middle of long bones. Yellow marrow provides a reserve energy source when other supplies of body fat have been reduced.

BLOOD CELL PRODUCTION

In addition to providing the body with mineral and fat reserves, bone marrow is also responsible for producing the blood cells needed to carry oxygen throughout the body and to fight deadly infections. These are produced in a specialized type of marrow known as *red marrow*, which is distinct from the previously mentioned yellow marrow. Among the cells produced by red marrow are the red blood cells, also known as *erythrocytes*. These contain the red pigment known as *hemoglobin*, which carries life-preserving oxygen from the lungs to the body's cells and tissues. Red marrow is also the source of the white blood cells, or *leukocytes,* that participate in the body's immune system and have the function of attacking disease-causing organisms that enter the body. *Platelets*, or *thrombocytes*, which are needed to form blood clots, are also manufactured in the red marrow. Moreover, if an individual becomes *anemic* (that is, if his or her supply of red blood cells falls to an abnormally low level), then yellow marrow can be transformed into red marrow in order to produce more red cells.

TYPES OF BONES

The human skeleton contains four different types of bones, each classified according to its shape.

- The *long bones* include most of the bones of the arms and legs, such as the ulna and radius of the forearm, and the tibia and fibula of the lower leg. These bones are longer than they are wide and have thin, cylindrical shafts.
- The *short bones* are about as long as they are wide, giving them something of a cube shape. The carpal (wrist) bones and tarsal (ankle) bones are examples of this type of bone.

- The *flat bones* are thin and broad. They include the bones of the roof of the skull, as well as the ribs and the sternum, or breastbone.
- The *irregular bones* have a number of different shapes. They include the *vertebral bones* of the spinal column, some of the bones of the skull, and the pelvic girdle (hipbone).

Some experts further classify certain bones as belonging to a fifth group, called *round* or *sesamoid bones*. These bones, usually small, are often found within tendons. They are normally located in the portion of the tendon that becomes compressed during movement. The kneecap, also known as the patella, is one example of a sesamoid bone.

A cross section of the femur, which runs through the upper leg from the hip to the knee. The marrow cavity occupies the center of the main shaft. Surrounding the marrow cavity are regions of spongy bone. The growth line is the region in which calcified bone tissue gradually replaces the cartilage that constitutes the early skeleton of the developing embryo.

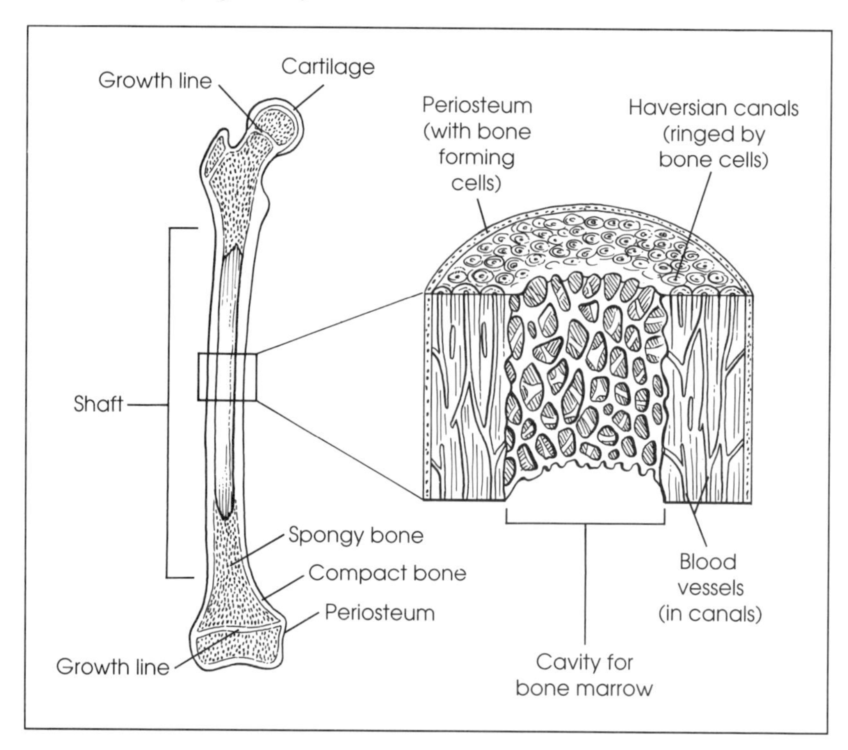

BONE STRUCTURE

The long bones are divided into sections. The long, thin shaft that runs between the ends of the bone is known as the *diaphysis*, while the wider section at each end of the bone is called the *epiphysis*. The epiphysis of each long bone meets the end of the next adjoining bone in a joint, which permits the two bones—and the respective body structures in which the bones are located—to move relative to one another. The joint can operate smoothly, without friction between the ends of the bones, because the epiphysis is covered by a layer of *cartilage*, a tough, resilient tissue containing strong fibers of the protein *collagen*. (Collagen is also found within the bone matrix, to which it imparts strength and flexibility.)

Except for the portion covered by cartilage, the long bones are enclosed by the *periosteum*, a durable layer of fibrous tissue (tissue composed of fibers), which contains blood vessels and connects with the tendons that link the bones to various muscles, and with the *ligaments* that link two bones that form a joint with one another. The functions of the periosteum also include helping to produce and repair bones.

Compact Bone

The walls of the diaphysis of a long bone consist of strong *compact bone*. The medullary cavity, containing the bone marrow, runs lengthwise through the middle of the diaphysis, surrounded by this compact bone.

Compact bone is made up of concentric rings or layers of dense bone surrounding hollow channels known as Haversian canals, with each of these rings of bone being called a lamella. Many Haversian canals run lengthwise throughout compact bone, where they carry nerves and blood vessels. Additionally, small holes, known as lacunae, occur within the lamellae and are connected by their own network of smaller horizontal canals known as canaliculi, which extend outward from each lacuna in raylike patterns. The bone matrix forms the walls of these different openings and passages.

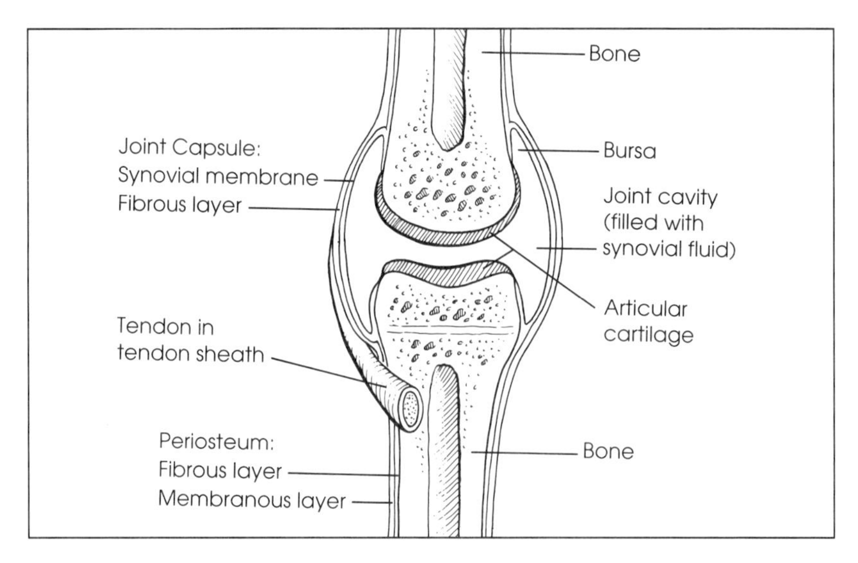

The structure of a synovial joint. Cartilage surrounding the ends of the bones and synovial fluid in the joint cavity cushion the shock when the bones come together during movement.

The bone cells known as *osteocytes*, which are the most common cells in mature bone and have the role of storing and releasing calcium, exist in the lacunae. The canaliculi connecting the lacunae carry nutrients and waste products from the blood vessels in the Haversian canals to the osteocytes, and vice versa.

Spongy Bone

Although the surface of the epiphysis consists of compact bone, its inner portion is composed of *spongy* or cancellous bone. Like compact bone, this spongy bone exists in layers made of the same cellular and matrix material as compact bone. Rather than being arranged in concentric circles, however, the layers of bone tissue in spongy bone run in various directions, forming spikelike structures known as trabeculae, which provide support against the forces—both external and internal—that act on the epiphysis. Additionally, spongy bone contains fewer Haversian canals than compact bone. The spongy bone contains red marrow. Such spongy bone, surrounded by a protective layer of compact bone, is also the major component of short, flat, and irregular bones.

JOINT STRUCTURE

Technically, a joint is an intersection between two bones. The human body contains three types of joints: synovial, fibrous, and cartilaginous.

Synovial Joints

Synovial joints are the only kinds of joints that can move freely. As discussed earlier, the end of each bone that forms such a joint is covered by a layer of cartilage known as *hyaline cartilage*. The smooth, pearly cushion provided by this cartilage can absorb the shock of running, jumping, and other activities while allowing the bones of the joint to slide smoothly against one another. A thin layer of synovial fluid nourishes and lubricates the synovial cartilage. Although the bones in some synovial joints contact each other directly, the bones in other synovial joints are separated from one another by a padlike articular disk, composed of cartilage, which provides further cushioning for the bone ends.

Holding the bones of a synovial joint together is the joint capsule, a structure whose outer layer consists primarily of dense, flexible connective tissue. The inner layer is made up of a thin synovial membrane. The capsule entirely encloses the joint, forming a saclike structure that extends from one of the bones of the joint to the other. In addition to the joint capsule, the bones of synovial joints are also held together by ligaments.

- A pivot joint contains a cylindrical region of bone that rests in a ring-shaped structure composed of bone and ligament. The only movement at this joint is the rotation of the central bone within the surrounding ring. An example of such a joint is the atlantoaxial joint that links the skull to the top of the backbone and allows the head to turn from side to side.
- A gliding joint is formed by two bone surfaces that are almost flat or only slightly curved. The bones in this joint can slide only from side to side or back and forth. Gliding joints can be found within the wrist and ankle.
- A ball-and-socket joint contains a bone with a round head that fits into a cuplike depression in another bone.

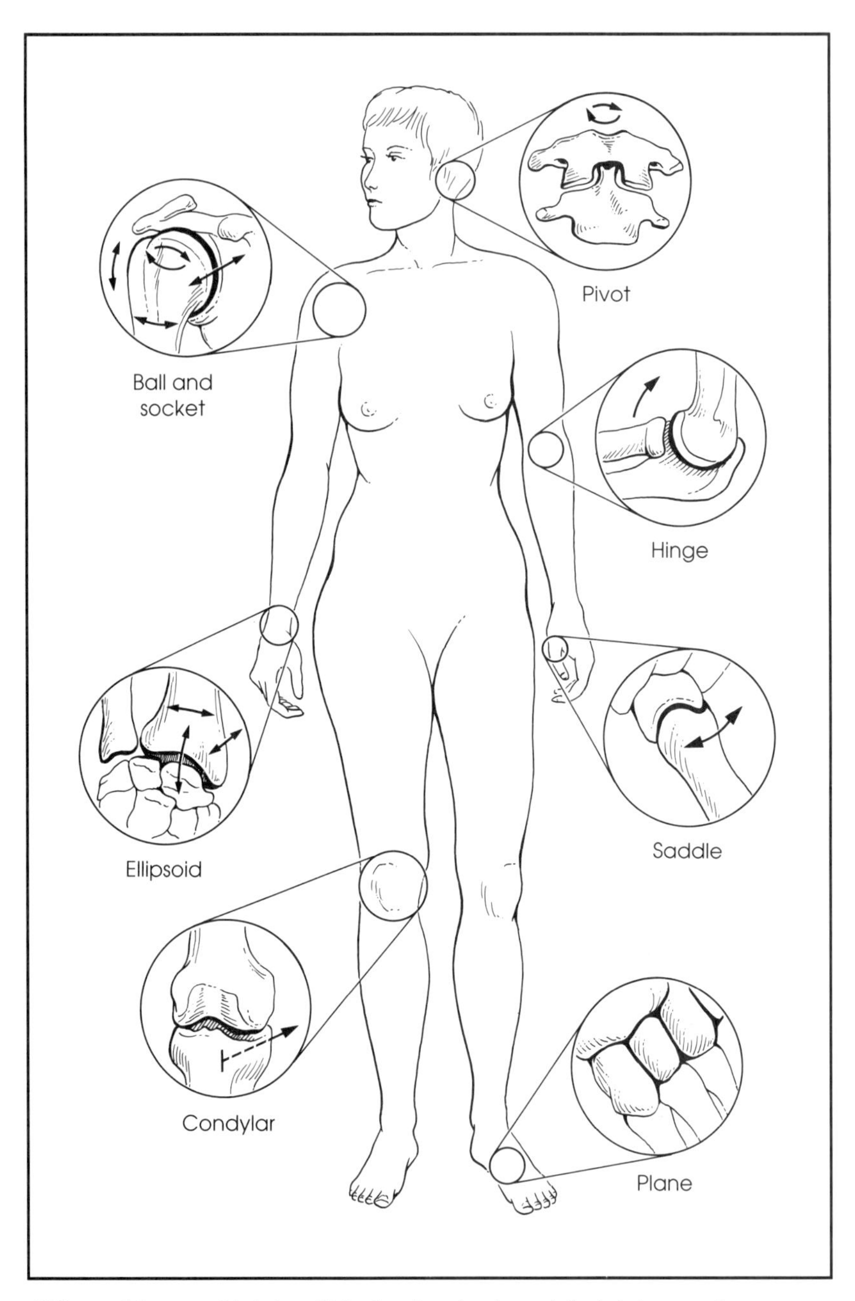

Different types of joints within the body. A saddle joint permits movement in two planes, such as from side to side and forward and backward. A hinge joint permits movement in only one plane, such as upward and downward. The bones of a gliding joint have nearly flat surfaces and slide across each other. In a ball-and-socket joint, the rounded surface of one bone fits into the cuplike end of another. In a condyloid joint, the semiround end of one bone moves within a flattened cup at the end of its neighbor. A pivot joint permits one bone to rotate around another.

It is this type of joint that links the upper bone of the arm with the shoulder and the upper bone of each leg with the hip. The ball-and-socket joint allows many different forms of movement.

- A saddle joint is formed between two saddle-shaped bone surfaces. This type of joint allows movement in several directions. The joint at the base of the thumb, formed between the carpal and metacarpal bones, is one example of a saddle joint.
- A condyloid joint looks like a modified version of the ball-and-socket joint. Ligaments and muscles at the joint, however, prevent the bones from rotating, although several other types of movement are possible. Examples of condyloid joints are the knuckles of the hands, which link the bones known as metacarpal bones to those known as the proximal phalanges of the fingers.
- A hinge joint is formed between the *convex* surface of one bone (that is, a surface that curves outward) and the *concave* surface of another (a surface that curves inward). This type of joint allows no rotation, permitting movement only in one axis. An example is the elbow, which basically permits the forearm to move only in an up-and-down direction.

Arthritis—an umbrella term for almost 125 different conditions resulting in joint pain—primarily attacks the synovial joints. This disease will be discussed in chapter 7.

Fibrous Joints

Bones in fibrous joints are bound tightly together by fibrous connective tissue. These joints allow little or, in most cases, no movement between the bones. There are three types of fibrous joints.

- The flat bones of the skull meet at an immovable fibrous joint known as a *suture*. Found only in the skull, this joint includes a thin sutural ligament that binds the bones together. Some of these ligaments are eventually transformed into bone.
- A syndesmosis is a joint held together by a long-fibered ligament called an interosseous ligament. This type of

joint exists, for example, between the distal ends of the tibia and fibula (the bones of the lower leg), where these bones meet to form the ankle. The ligament in a syndesmosis is flexible, which means that some movement of the two bones is possible.

- A gomphosis resembles a plug in a socket. This type of joint occurs where the root of a tooth attaches to the jawbone, to which it is bound by a periodontal ligament. This is also an immovable joint.

Cartilaginous Joints

Cartilaginous joints are bound either by fibrocartilage or hyaline cartilage. Fibrocartilage is made up of thick bundles of tough white collagen fibers. There are two types of cartilaginous joints.

- A synchondrosis, in which the bones are connected by hyaline cartilage, is often a temporary joint that vanishes as the body matures. The epiphyseal plate that separates the epiphyses of the growing long bones from the remainder of the bone is one example of a synchondrosis.
- A symphysis consists of a thin layer of hyaline cartilage covering the edges of two or more bones that meet to form a joint. Underneath the hyaline cartilage is a pad of fibrocartilage. Because the fibrocartilage is resilient, a symphysis is capable of some expansion and flexion. The cartilaginous intervertebral disks that separate the vertebral bones of the spine are an example of this type of joint. Each of these disks consists of a band of fibrocartilage surrounding a core that has the texture of gelatin. Because each disk is both compressible and flexible, all of the intervertebral disks, acting together, allow the back to twist and bend.

CHAPTER 3

BONE GROWTH AND DEVELOPMENT

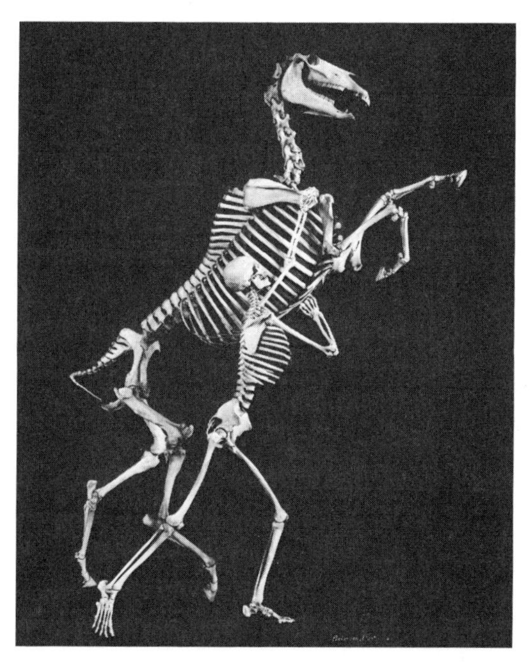

A comparison of the skeleton of a human being and a horse reveals similarities in skeletal structures between different species of mammals.

Two types of bone development occur during fetal development. The simplest type is called intramembranous ossification. The other is endochondral ossification.

INTRAMEMBRANOUS OSSIFICATION

Relatively few bones are formed by intramembranous ossification. They include the flat bones of the roof of the skull and some of the facial bones. Such ossification begins with the development of layers of crude connective tissue laced with blood vessels. The connective tissue cells surrounding the blood vessels enlarge and transform into specialized bone cells known as *osteoblasts*, which are responsible for manufacturing the bone matrix. This process produces spongy bone. Eventually, lacunae form in the spongy bone, and the osteoblasts trapped inside these lacunae develop into *osteocytes*.

The original connective tissue cells on the outside of this developing bone produce the periosteum. The osteoblasts inside the periosteum form a layer of compact bone that covers the spongy bone.

ENDOCHONDRAL OSSIFICATION

Most bones develop through the process of endochondral ossification. In this process, hyaline cartilage in the growing fetus gradually takes a shape similar to that of the completed bone. The cartilage is gradually surrounded by a membrane in the spongy bone, the perichondrium, which consists of fibrous connective tissue. As the hyaline cartilage within the perichondrium begins to break down, the perichondrium develops into a periosteum. Some of the connective tissue cells of this covering become osteoblasts, and these form the layer of compact bone that surrounds endochondral bone.

In the formation of long bones, cartilage cells in the diaphysis enlarge and compress the protein substance known as intercellular material, forcing it to thin out. This intercellular protein then begins to calcify, or harden, as calcium is deposited in it, blocking the access of nutrients that would normally feed the cartilage cells. As a result, the cartilage dies, leaving holes in the calcified material. Blood vessels from the periosteum then invade these gaps. These vessels carry connective tissue cells that travel through the blood from the periosteum to the holes in the calcified intercellular material, where some of these cells are transformed into osteoblasts. These cells then begin to form spongy bone, starting at the center of the diaphysis in an area called the primary ossification center. Eventually, after it has already been laid down, this spongy bone is *resorbed* by the osteoclasts, creating the medullary cavity at the center of the bone.

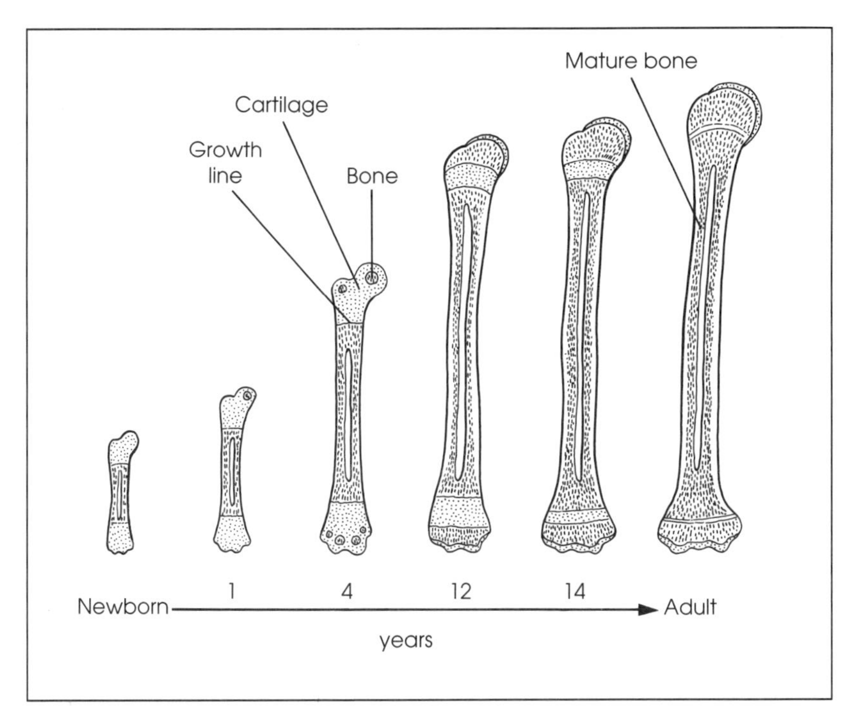

In the process known as endochondral ossification, an early "model" of each bone, formed of cartilage during the growth of the human embryo, becomes enlarged and transformed into true bone by the age of 25.

By the time an individual is born, the diaphyses of the long bones have already been formed. The epiphyses of a newborn infant are, however, still composed of cartilage, although the matrices of the epiphyses soon calcify and a secondary ossification center forms in each epiphysis, creating spongy bone. However, a section of cartilage, the epiphyseal cartilage or plate, remains between each epiphysis and the diaphysis that forms the shaft of the bone. Additionally, a thin layer of cartilage remains on the surface of each epiphysis, where it cushions the joints.

In a newborn, the marrow cavities of the bones contain only red marrow. By the time an individual reaches adulthood, however, yellow marrow has taken over these cavities, and red marrow is found only in certain regions of spongy bone, including the vertebrae and, as mentioned, some epiphyses.

ENDOCHONDRAL BONE GROWTH

Throughout childhood and adolescence, the epiphyseal plates allow the long bones to continue to increase in length. Within each plate are four layers of cartilage cells, each serving a different function. The first layer contains resting cells that connect the epiphyseal plate to the epiphysis of the bone. The second layer consists of cartilage cells undergoing division, forming new cells that cause the plate to grow longer. A third layer is built of older cells that enlarge, causing the bone to lengthen even more. Calcium salts invade this third layer, and the oldest cells die, forming the fourth layer of the plate. The dead cells are resorbed, and osteoblasts, which infiltrate the area closest to the diaphysis, replace them with bone.

By the time of late adolescence, the rate of cartilage growth becomes slower than the rate of bone formation. As a result, by the time a woman is in her early twenties and a man is in his mid-twenties, the cartilage in the epiphyseal plate has been completely replaced by bone. This means that the long bone can no longer increase in length. An epiphyseal line remains, however, showing where the cartilaginous plate had been.

Even though the disappearance of the epiphyseal plate means that the bone cannot become longer, the diameter of the diaphysis can increase. This results when *osteogenic cells* in the inner layer of the periosteum change into osteoblasts, which deposit more bone. While this is happening, older layers of compact bone are absorbed by osteoclasts. As a result, the medullary cavity becomes wider, although the compact bone itself may become only slightly thicker.

FACTORS INFLUENCING BONE GROWTH

Stress

Physical stress on a bone can encourage the bone to grow and become stronger. As force is applied to the bone, more collagen fibers and inorganic salts are deposited in the bone matrix. Additionally, the collagen fibers of the bone can change their position when necessary,

The stress of gravity is an essential factor in normal bone growth. Loss of bone has occurred during weightlessness in space travel and under other circumstances in which gravity and other forces cease to act upon the body.

in order to provide maximum support against repeated stress in a particular direction.

Body weight is one of the main forces acting on the skeleton, and the heavier a person is, the stronger his or her bones must be to support this weight. This means that an obese person must typically have a heavy skeleton. Functional forces also exert an important influence on the bones. An example of such a force is the force exerted as muscles attached to the skeleton contract and pull on it. Weight lifters develop a skeleton thick enough to prevent their strengthened muscles from literally snapping the bones to which these muscles are connected.

Just as stress can make bones stronger, a lack of it can cause them to weaken, because inorganic salts will be withdrawn from the bone matrix. When an arm or leg is immobilized in a cast, for example, bones can *atrophy* (decrease in size as the tissue wastes away). This can also happen to astronauts in the near weightlessness of space.

Hormones

Bone growth is also influenced by a number of hormones produced by the body, including those secreted by the pituitary, thyroid, and parathyroid glands, as well as those from the ovaries (female sex organs)

A giant and a dwarf. Gigantism results from the oversecretion of human growth hormone by the pituitary gland, extending the period of bone growth beyond its normal range. In dwarfism, the pituitary gland stops secreting growth hormone prematurely.

and testes (male sex organs). For example, an increased concentration of parathormone, a hormone secreted by the parathyroid gland, encourages osteoclasts to resorb bone. In contrast, calcitonin, secreted by the thyroid gland, causes osteoclasts to absorb less bone and may actually encourage bone formation. If too much of this hormone is secreted, the epiphyseal plate may be transformed completely into bone too early in an individual's development, halting normal growth.

Human growth hormone, secreted by the pituitary gland, stimulates the generation of cartilage cells in the epiphyseal plates. A child who fails to produce enough of this hormone can become a *pituitary dwarf*, a condition in which the body is normally proportioned but extremely short. Excessive secretion of growth hormone, on the other hand, can turn a person into a *pituitary giant*.

Nutrition

Good nutrition is essential to proper bone growth. The body needs vitamin D, for example, in order to absorb calcium from the gastrointestinal tract into the bloodstream. Calcium, as mentioned, is an important element in the formation of the bone matrix. If the body contains too little of this mineral, it can develop the bone-deforming disease known as *rickets* in children and *osteomalacia* in adults.

Vitamin A is needed for the resorption process that takes place during bone development, and a deficiency of this vitamin may therefore interfere with normal development. Vitamin C plays a role in the production of collagen and, if supplied in insufficient quantities, can make bones excessively slender and breakable.

Phosphorus discourages bone formation. In the form of the chemical compounds known as phosphates, in which it is combined with oxygen, it joins with calcium, interfering with the ability of vitamin D to help the body absorb calcium from the diet. This means that when the concentration of phosphate in the blood rises, the concentration of calcium falls. Many foods in the modern American diet are high in phosphorus, including meat, fish, and fowl, as well as soft drinks, processed foods, and snack foods, and the average individual may consume twice as much phosphorus as calcium.

On the other hand, hormones acting on the bones help to regulate the concentration of phosphate in the blood and elsewhere in the body. Parathyroid hormone, for example, releases phosphate from the bones just as it promotes the release of calcium. This may seem like a contradiction, since the effects of these two substances oppose one another, with increased phosphate levels in the blood being accompanied by decreased calcium levels, and vice versa. However, parathyroid hormone also increases the excretion of phosphate in the urine and therefore ultimately decreases the amount of phosphate in the blood.

BONES AND AGING

With aging, an individual loses bone mass. The amount of bone lost, however, differs with gender and race. For example, men tend to lose less bone than women. Moreover, there is significantly less bone loss among blacks than among whites in the United States.

Factors Related to Bone Loss

Several factors appear to be related to greater bone loss among women than among men. One important factor is *menopause*, during which women stop menstruating. Following menopause, a woman produces less of the hormone *estrogen* than during her previous adult years. Because estrogen can dampen the bone-resorbing effects of parathyroid hormone, these effects can increase when a woman's estrogen level decreases.

Even before menopause, gradual bone loss is common among women, in whom the calcium content of bones undergoes a steady reduction after age 40. This means that by the time a woman reaches age 70, her skeleton may have lost 28% of its calcium. Ultimately, women lose about 35% of the compact bone around their diaphyses, and about 50% of their spongy bone. Men, on the other hand, usually do not begin to lose bone calcium until they are past age 60. In contrast to women, men lose only about 25% of their cortical bone and a bit more than 30% of their trabecular bone. Moreover, even before men or women begin to lose calcium, the male skeleton normally has higher levels of calcium than does its female counterpart.

The reasons for the differences in calcium loss between men and women are not fully understood, but scientists believe that other factors in addition to menopause contribute to it. For example, women tend to be more weight conscious than men and so tend to avoid high-fat dairy foods. As a result, they may miss out on the substantial quantities of calcium in these foods.

In the same way, several factors also appear to be related to the greater bone loss among whites than blacks. Although it is not known why, black people tend to have 5% to 10% more mineral content in their bones than do white people. Blacks also appear to have a lower rate of bone "turnover" than whites, with both the formation of new bone and the resorption of old bone occurring more slowly than in whites.

CHAPTER 4

THE MUSCULAR SYSTEM

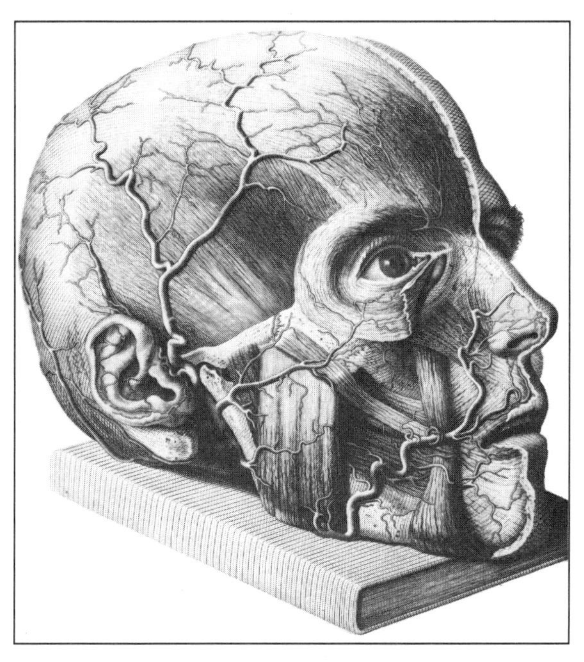

This drawing, made in 1801 by Antonio Scarpa, shows the highly complex muscle structure that enables people to control their jaw movements and make complex facial expressions.

Muscles produce movement. This movement can range from a voluntary motion of the arms and legs to the involuntary contraction of the intestines that moves food through the digestive tract. Breathing, too, depends on muscle—in this case the *diaphragm*, a wall of muscle whose regular contractions are responsible for the expansion of the lungs and the inhalation of air, and whose relaxations compress the lungs and force them to exhale the air they have inhaled.

The body contains three types of muscles: *skeletal*, *smooth*, and *cardiac*. The structure and function of each will be described in detail, although skeletal muscle, because of its intimate relationship with the body's bone structure, will be the primary form of muscle tissue discussed throughout the text.

Some of the outer skeletal muscles, fascia, tendons, and ligaments of the body.

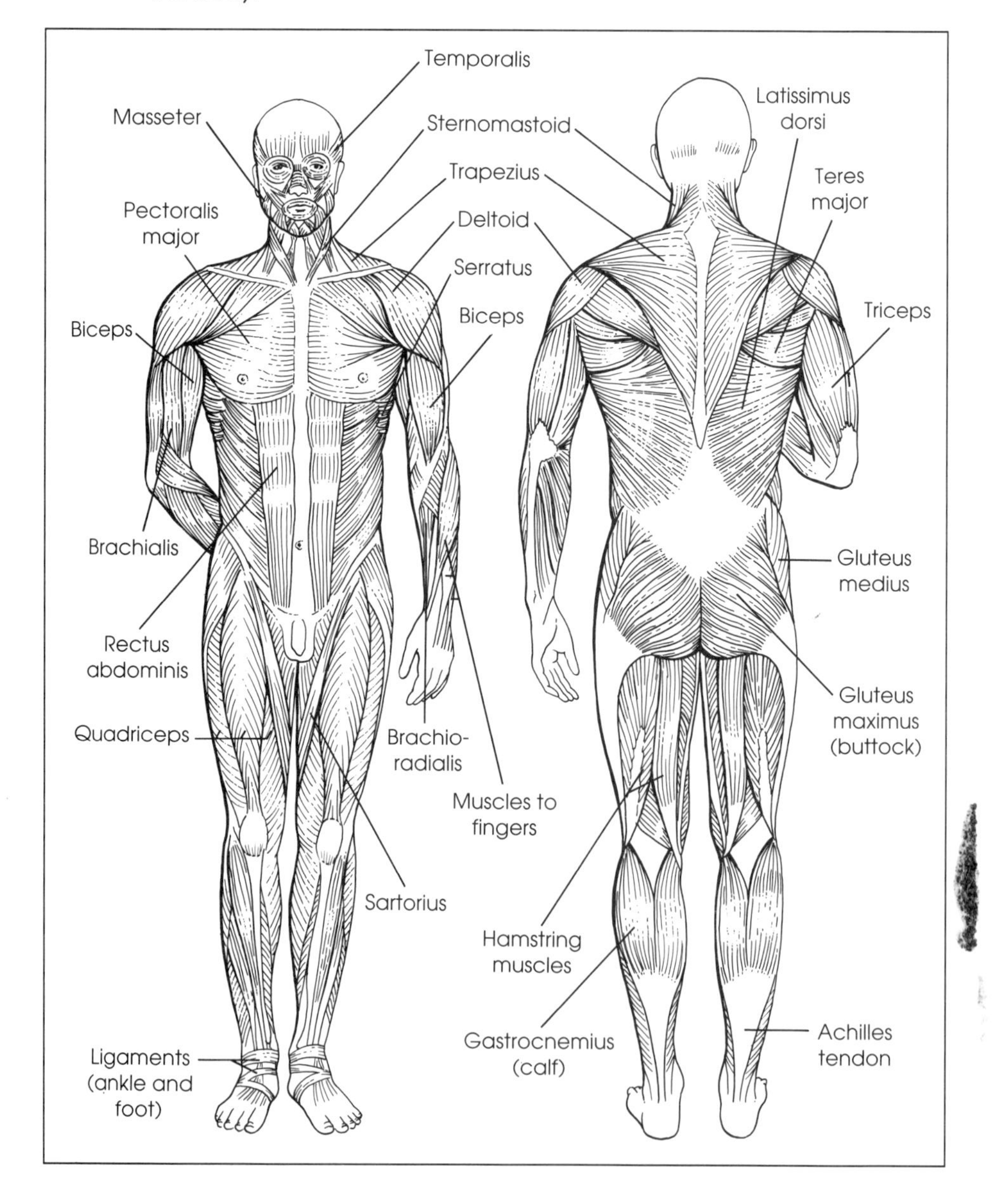

SKELETAL MUSCLE

Skeletal muscles are attached to bones and are responsible for the three classes of lever movements discussed in chapter 2. A skeletal muscle arises at its *origin*—the site at which it is attached to a bone that does not move. Its *insertion* is the point of its attachment to the bone that the muscle moves. As the most numerous muscles in the body, skeletal muscles constitute about 40% to 50% of an individual's total body weight. In contrast to the body's smooth or cardiac muscles, the skeletal muscles are involved in voluntary movements.

Skeletal muscles are composed of long, cylindrically shaped cells called fibers, which are rounded at each end. A layer of connective tissue, called the perimysium, winds through the muscle, surrounding groups of these muscle fibers and separating them into individual bundles called *fascicles*. Each individual muscle fiber is in turn covered with a thin layer of connective tissue called the endomysium. When the fascia extends beyond the ends of the muscle, it forms the tendons that connect with the periosteum of a bone, attaching the muscle to the

Cross-sectional view of the structure of a skeletal muscle. The muscle is composed of many fascicles, which are in turn composed of individual cylindrically shaped muscle fibers.

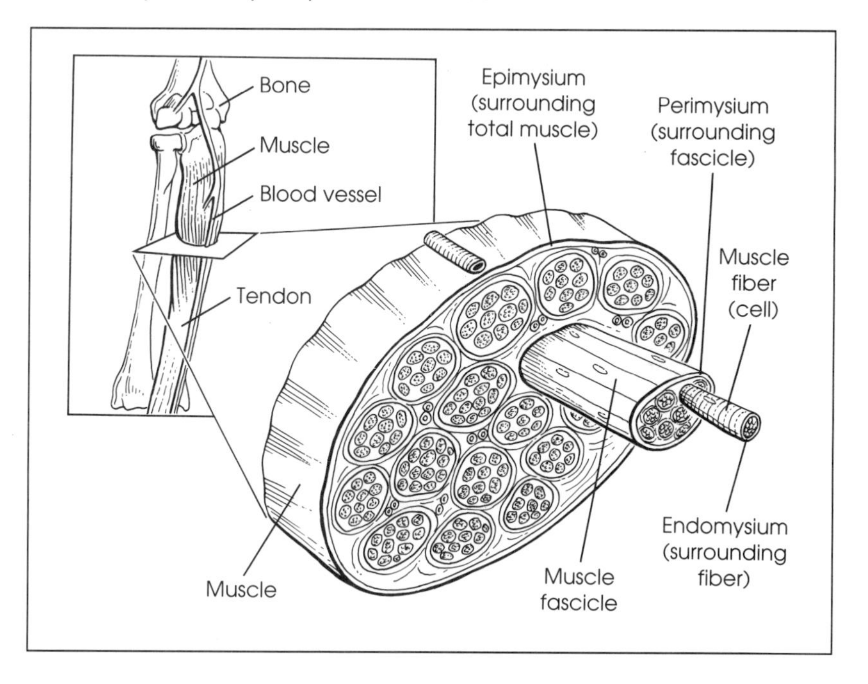

bone, although tendons may also be connected to the skin—as in the case of the facial muscles—and even to other muscles. If two or more muscles work in combination, their fascia can form sheets known as aponeuroses, which bind the cooperating muscles together.

Both the perimysium and the endomysium are derived from a larger layer of connective tissue called the epimysium, which surrounds the entire muscle and is itself surrounded by a sheet of connective tissue known as the *fascia*, which protects the muscle and contains nerves, blood vessels, and other structures that serve the muscle.

Fibers

An unusual feature of the long, cylindrical cells of skeletal muscle is that each contains several nuclei instead of the single nucleus found in the cells of most other kinds of tissue. The cell membrane, which directly surrounds the fiber and is distinct from the overlying endomysium, is called the *sarcolemma*. The cell substance, or cytoplasm, which is contained within the membrane of the muscle cell, is specifically called the *sarcoplasm*. Within it are a series of fibers known as *myofibrils* as well as a network of small, fluid-filled sacs and small canals known as the *sarcoplasmic reticulum*.

It is the myofibrillae that are responsible for the movement of muscles. They contain thick strands of a type of protein known as *myosin* and thin threads of another protein known as *actin*. When examined under the microscope, these protein myofilaments form light and dark bands that are unique to muscle fibers. The dark bands, called *A bands*, contain mostly myosin, while the light bands, called *I bands*, consist predominantly of actin. The roles of actin and myosin in muscle contraction will be discussed later in this chapter.

Motor Neurons

The movement of muscles is controlled by *motor neurons*, specialized nerve cells that stimulate muscle contraction. These cells are parts of the motor nerves that extend from the brain or spinal cord and ultimately branch into muscle fibers, each of which connects to an individual fiber within one of the body's skeletal muscles.

The intersection of a nerve and muscle fiber is called the *neuromuscular junction.* At this junction, the sarcolemma, or muscle cell membrane, is modified into what is known as a *motor end plate*. Here the cell membrane of the muscle cell contains recesses and is also intricately folded within each recess. The nerve fiber divides into branches that extend into these recesses. When an impulse is transmitted through the nerve fiber, chemical substances called *neurotransmitters* are released from the ends of the branches of the nerve fiber and stimulate the muscle fiber to contract.

The number of muscle fibers served by a single motor neuron varies from one part of the body to another and depends upon how delicate the movements must be in a particular area. Motor neurons controlling the most delicate muscle contractions have fewer nerve fibers and therefore attach to fewer muscle fibers than neurons controlling muscles that do not need to operate at quite so fine a level. Thus, fewer than 10 muscle fibers can be linked to a neuron controlling the delicate movements of the eye, whereas a nerve running to a large muscle in the back may connect to more than a hundred muscle fibers. A motor neuron and the muscle cells to which its fibers connect are known collectively as a *motor unit.*

MUSCLE CONTRACTION

Muscle contractions are produced by a complex series of chemical reactions. When a nerve impulse reaches the motor end plate on a muscle fiber, the neurotransmitter known as *acetylcholine* transforms the impulse into a chemical signal that stimulates an electrical muscle impulse in the sarcolemmal membrane of the fiber. This impulse is transmitted through the sarcoplasm, stimulating the release of calcium ions within the sarcoplasmic reticulum. The calcium ions in turn cause the actin and myosin filaments in the muscle fiber to link together in a process that makes the muscle fiber contract. Precisely how actin and myosin produce this contraction is not fully understood, although the mechanisms suggested for its occurrence will be discussed more fully below. The simultaneous stimulation of large numbers of muscle fibers by a nerve impulse transmitted into the many branches of the nerve that are connected to the fibers is what makes an entire muscle contract.

Once a muscle has contracted, its muscle fibers quickly allow it to relax once again. Calcium ions are swept back into the sarcoplasmic reticulum, breaking the links between actin and myosin. Additionally, the *enzyme* known as *acetylcholinesterase* acts chemically to break down or digest the acetylcholine released at the end plate on each muscle fiber, ending the transmission of contraction impulses to the sarcolemmal membrane of the fiber.

ACTIN-MYOSIN ATTACHMENTS AND MUSCLE CONTRACTION

The Sliding-Filament Theory

Although the way in which actin and myosin bring about muscle contraction is not fully understood, the sliding-filament theory is one possible explanation. It suggests that stimulation of the fiber prompts the formation of tiny crossbridges that extend from the myosin filament and attach to active sites on the actin filament. The release of calcium ions within the muscle fiber exposes these active sites, facili-tating the attachment of the two kinds of fiber to one another. Each crossbridge exerts a pull on the actin filament, causing the actin and myosin filaments to slide past one another. Under the influence of chemical substances released in the binding process, each crossbridge is then disconnected from its binding site on the actin filament and moves to a neighboring binding site. Since the process happens simultaneously in all of the cells of a muscle, the entire muscle contracts.

A more recent theory suggests that actin and myosin interact much as they do in the sliding-filament theory, but rather than sliding past the actin filament as its crossbridges attach to one active site after another on the actin filament, myosin rotates like a corkscrew as it links up with the active sites on the actin filament. The resulting muscle contraction is, however, the same.

Troponin and Tropomyosin

In resting muscle, two different substances, troponin and tropomyosin, cover the active sites on the actin filaments of the muscle. Calcium ions released from the sarcoplasmic reticulum attach to troponin, forcing it to shift its position and in turn pulling the tropomyosin, to which the troponin is attached, away from the active sites, thereby exposing them.

FUEL INTO MOTION

The linkage of actin and myosin filaments that leads to muscle contraction depends upon energy provided by the chemical compound known as *adenosine triphosphate*, or ATP, which is present in muscle fibers. This compound contains three units of the substance known as phosphate (PO_4), each of which is bound into the ATP molecule by a high-energy bond. When such a bond is broken and a phosphate unit is released, a great deal of energy is given off, and it is this energy that powers the actin-myosin interaction. The ATP molecule, now left with only two phosphates, becomes what is called *adenosine diphosphate*, or ADP. The ATP that energizes the linkage of myosin to actin and the ensuing muscle contraction becomes temporarily bound to the myosin crossbridge. It is the myosin itself, acting in the capacity of an enzyme known as myosin adenosine triphosphatase, or myosin ATPase, that breaks the high-energy bond of the ATP to release the energy needed for the reaction.

The ATP that energizes muscle contraction is manufactured by structures known as *mitochondria*, located within the muscle fiber. However, the ATP supply is limited and can fuel muscle contraction for only a short time before it becomes used up. As a result, the muscle fiber must be able to transform ADP back into ATP so that more energy can be produced. That means it must come up with another energy source for reconnecting, in the form of high-energy bonds, the loose phosphate units released in the original bond-breaking process to the ADP that was left after the bond breaking.

Creatine Phosphate

One source of energy for re-creating ATP is creatine phosphate, a molecule that also contains phosphate units connected via high-energy bonds. When the supply of ATP runs low, the phosphate bonds in creatine phosphate break, releasing energy that the muscle fiber can use to transform ADP back into ATP in preparation for the next interaction of myosin and actin. The resulting ATP is then used once again to supply the energy needed for muscle contraction.

When a muscle is working its hardest, however, there is only enough ATP and creatine phosphate to supply the energy needed for a few seconds of movement. To keep the muscle operating, the body turns to cellular respiration.

Cellular Respiration

When they are especially active, muscle fibers use the process of *cellular respiration* to turn glucose—a type of sugar—into fuel for energizing muscle contraction. The glucose molecules needed for this process are stored within the muscle fiber in the form of glycogen, a substance consisting of glucose molecules bound to one another, from which they can be released as needed.

In the first phase of cellular respiration, molecules of glucose, released from glycogen, are broken down in a process known as *glycolysis*. Because it does not use oxygen, this process is defined as *anaerobic*. In it, a single molecule of glucose is transformed into two molecules of the substance known as pyruvic acid, with the generation of four new molecules of ATP.

In the next phase of cellular respiration, called an *aerobic* reaction because it consumes oxygen, the pyruvic acid produced in the preceding phase combines with a substance known as coenzyme A to form a larger complex known as acetyl-coenzyme A (acetyl co-A). The latter substance then combines with a third substance—oxaloacetic acid—which is already present in the muscle fiber. In the next step of the process, this complex of acetyl co-A and oxaloacetic acid gives rise to the substance known as citric acid, which is the first component in a series of reactions known as the *citric acid cycle* (also known as the Krebs cycle, for the British biochemist Sir Hans Adolph Krebs (1900–81), who identified it as a vital pathway by which various kinds of mammalian cells produce energy). Every completion of the citric acid cycle generates two new molecules of ATP for the cell, as well as two molecules of carbon dioxide and eight atoms of the element hydrogen. But that is not the end of the matter, since the hydrogen atoms enter

Sir Hans Adolph Krebs, who identified the citric acid cycle as an important pathway by which muscle cells provide themselves with energy while replenishing their supply of the energy-containing substance known as adenosine triphosphate (ATP).

into yet another series of reactions known as the electron transport system, which employs specialized molecules known as cytochromes, as well as oxygen, to manufacture 32 additional molecules of ATP.

The oxygen that muscle cells need to carry out the citric acid cycle comes from the blood. The muscle cells store some of this oxygen by combining it with a pigment named *myoglobin*, which holds it in reserve for use as needed. This helps the muscle continue to function even when muscle contraction compresses and closes off some of the blood vessels that otherwise carry oxygen to the muscle.

EXERCISE AND OXYGEN DEBT

Although the body has an efficient system for supplying the muscles with oxygen during normal activities, strenuous muscle movement can quickly consume this oxygen—which is why exercise results in a rapid heartbeat and heavy breathing in an attempt to keep up the intake of oxygen and pump it to the cells. In a further effort to keep its muscles operating efficiently, the body also relies increasingly on energy provided by anaerobic respiration. Unfortunately, this compensatory effort cannot be maintained for very long.

Ultimately, an insufficient supply of oxygen spurs the transformation of pyruvic acid into *lactic acid*, a waste product of metabolism that gathers in the muscle fibers and makes muscle contraction increasingly difficult. As its concentration in the muscle cell increases, the lactic acid can fortunately diffuse out of the cell and into the bloodstream, which transports it to the liver.

The muscles of the body can be trained not only for strength but for endurance. Here, Geoff Smith wins the 26-mile 1984 Boston Marathon in just over 2 hours and 10 minutes.

Ordinarily, the liver can transform this lactic acid into glucose through the process of aerobic glycolysis. But in order to do this, the liver requires ATP, which is in short supply during any period of exertion, because the body is already using most of its ATP for the process of muscle contraction. As a result, lactic acid builds up in the liver, creating an *oxygen debt* that can be paid only when the body has produced a sufficient quantity of ATP to transform all of the lactic acid in the liver into glucose and return the overall ATP supply to its pre-exercise levels. (Creatine phosphate, which is also consumed during the period of exertion, must be returned to its previous levels as well.) Because the transformation of lactic acid to glucose occurs slowly, the oxygen debt following heavy or sustained exercise may last for several hours.

EXERCISE AND MUSCLE FATIGUE

Just how much lactic acid the muscles can tolerate before they lose the ability to contract varies from person to person. The fatigue results from lactic acid–induced chemical changes in the muscle fiber that prevent the cell from responding to nerve stimulation. Regular exercise can, however, reduce the amount of lactic acid the muscles produce, which means that athletes tend to tire less quickly than other people.

FAST-TWITCH AND SLOW-TWITCH MUSCLE FIBERS

Skeletal muscles contain two types of fibers: *fast-twitch* and *slow-twitch* fibers. The performance of a specific muscle is linked to the kind of fiber that is predominant in that muscle.

Fast-Twitch Fibers

Although fast-twitch fibers use ATP more rapidly than do slow-twitch fibers, they have a high content of glycogen, a relatively low supply of myoglobin, and contain small mitochondria, and they are therefore better suited to obtaining ATP from anaerobic respiration than from the more efficient aerobic cycle. They are therefore best for performing rapid contractions over a short period and tend to tire quickly.

Slow-Twitch Fibers

As their name implies, slow-twitch fibers contract more slowly than do fast-twitch fibers. They also receive a better blood supply and contain more mitochondria and a greater quantity of myoglobin. In addition, slow-twitch fibers break down ATP at a slower rate than fast-twitch fibers, which means that during contraction, they can generate this energy-supplying substance faster than they use it up. For all of these reasons, slow-twitch fibers are best suited to aerobic respiration.

With regard to the varying proportions in which skeletal muscles contain fast- and slow-twitch fibers, the *long muscles* of the back, which are needed to maintain posture, consist primarily of slow-twitch fibers. Because their strong blood supply and preponderance of myoglobin give them a reddish color, such muscles are called *red muscles*. By contrast, some muscles of the hand, because of their smaller blood supply and the more limited quantity of myoglobin they contain, are called *white muscles*. These muscles consist mainly of fast-twitch fibers.

The percentage of fast-twitch and slow-twitch fibers in the body can affect one's athletic abilities. Some persons who have a majority of fast-twitch fibers in their legs can run short distances very quickly and tend to be particularly good sprinters. Others, with a higher percentage of slow-twitch fibers, often excel at long-distance running. In some cases, an individual has a fairly even balance of both types of fibers and is adept at both aerobic and anaerobic activities.

SKELETAL MUSCLES AND AGING

After the age of 40, the skeletal muscles tend to lose mass and strength. One change that affects these muscles appears to be a decrease in the body's supply of human growth hormone, which, besides being essential to growth, helps the body to burn fat. The suspected decrease in growth hormone with age may lead to an increase in the body's fat content and a decrease in its content of muscle tissue. The decrease in muscle appears to be accompanied by a decline in the number of muscle motor units—the combinations of a motor neuron with a muscle cell. Furthermore, the sarcoplasmic reticulum enlarges with age, a change that apparently increases the rate at which it resorbs calcium ions. If the reticulum begins to resorb these ions more quickly than they can perform their function of facilitating muscle contraction, the body's muscle contraction may suffer. Researchers have also found that as aging occurs, the energy-releasing enzyme activity of myosin is reduced. Despite the tissue deterioration that occurs with aging, muscles can apparently withstand these changes. As a result, even elderly muscles can perform well, assuming that a senior citizen is in good health and exercises regularly.

In a 1990 study, researchers at the University of Virginia suggested that it may be possible to improve muscle strength in senior citizens by using injections of synthetic human growth hormone. The study found that men between the ages of 61 and 81 who took hormone injections three times per week had an 8.8% increase in their lean body mass—the weight of the body without any fat, which includes muscle—as well as a 7.1% increase in skin thickness and a 14.4% decrease in fatty tissue. Unfortunately, using large doses of growth hormone, or taking it for an extended time, has been linked to high blood pressure, diabetes, and heart disease. More research is therefore needed to determine whether such treatments can produce lasting improvements in skeletal muscle.

SKELETAL MUSCLE SHAPE AND FUNCTION

The shapes of muscle vary according to their function. Circular, or sphincter, muscles are responsible for the opening and closing of openings in the body. The orbicularis orbis muscle that surrounds the mouth and the orbicularis oculi muscle that encircles the eye and has the role of closing the eyelid are both sphincter muscles. Other sphincter muscles control the flow of partly digested food from the stomach into the small intestine, while still others, at the outlets of the urinary and digestive tracts, permit voluntary control of the flow of wastes from the body.

Some muscles, whose fibers are parallel to one another, have a straplike or quadrate shape. The sternohyoid muscle, connecting the top of the breastbone with the hyoid bone in the upper front of the neck, is one such straplike muscle. By contrast, the biceps muscle, which runs through the upper arm and whose contractions raise the lower arm, is what is known as a fusiform muscle, with a thick central region or belly that tapers off at both ends, where its tendons are located.

Both the muscles themselves and the fascia that connect them to bones also exhibit other differences in shape. Some muscles have many

Shapes of skeletal muscles: 1) unipennate; 2) bipennate; 3) multipennate; 4) rhomboid; 5) trapezoid; 6) quadrate; 7) fusiform; 8) digastric; 9) bicipital.

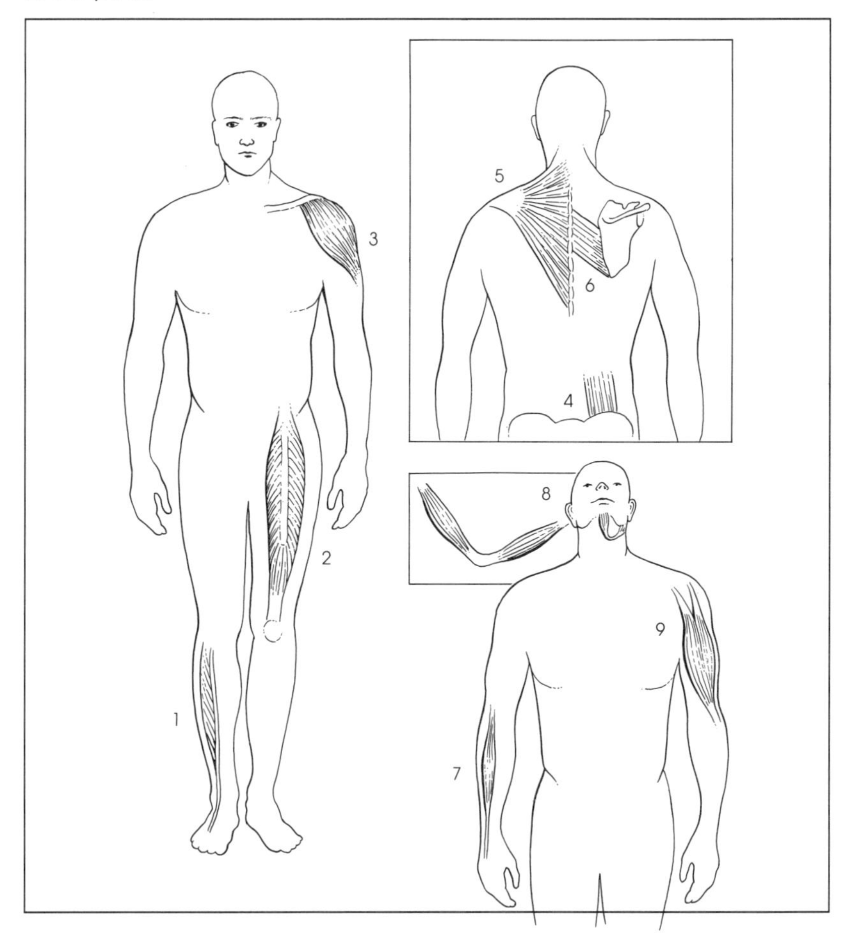

fascicles that assume a fanlike shape. Such is the case with the trapezius muscle that connects the shoulder blades with the back of the skull and upper spinal column. A muscle of this kind is called a *multipennate* muscle. By comparison, the fascicles of some other muscles begin as a close group on one bone, but then angle outward so that the fascicles connect along the length of another bone. A muscle of this kind is known as a *unipennate* muscle. A third type of muscle, known as a *bipennate* muscle, has fascicles that are positioned in a line along a single bone, but then fan outward in opposite directions from one another, connecting the first bone in a lengthwise manner to one or two other bones.

SMOOTH MUSCLE

Smooth muscle is responsible for many of the body's automatic, or involuntary, muscle movements. This type of muscle is found in the walls of the stomach and along the intestines, where its contractions push food through the digestive system in the process known as *peristalsis*. Smooth muscle is also a major structural component in the walls of the blood vessels known as *arteries*, which carry blood outward from the heart to the body's tissues, and a lesser component of the walls of the vessels known as *veins*, which carry blood back to the heart. Within the walls of the small arteries known as arterioles, smooth muscles play a role in controlling blood pressure, which increases as these muscles contract and bring about a narrowing of the arterioles. Conversely, blood pressure decreases when the muscles relax and the vessels widen.

Smooth muscle derives its name from its lack of the dark and light bands found in skeletal muscle. Although smooth muscle cells, like those of skeletal muscle, do contain actin and myosin, these proteins are not arranged in the same pattern as the actin and myosin of skeletal muscle cells.

Another difference between smooth and skeletal muscle is that smooth muscle is not attached to bones. Nor are its fibers as long as skeletal muscle cells, and rather than containing several cell nuclei, as do the cells of skeletal muscle, they contain only one nucleus. Moreover, the sarcoplasmic reticulum of the smooth muscle cell is not as well developed as that of skeletal muscle, and the composition of the myosin in these cells is somewhat different from that in skeletal muscle. And although the basic mechanism for smooth muscle contraction is similar to that for skeletal muscle fibers, it does vary somewhat from the latter. Thus, rather than containing troponin, smooth muscle cells carry a protein called calmodulin, which serves a similar function, binding with calcium and initiating muscle contraction.

In most cases, the cells in smooth muscle tissue are connected to one another in such a way that when a nerve fiber stimulates one cell to contract, the muscle fiber transmits the signal to other smooth muscle cells in a domino effect that causes the other cells to contract. The cell that receives the initial nerve stimulation is called a pacemaker cell.

Not all smooth muscle tissue is structured to produce a domino effect. The smooth muscle fibers in some organs are not connected to one another and must be stimulated individually, by separate nerve endings. Such is the case in the eye, where muscles must function quickly to bring images into proper focus.

Enlarged views of the structures of skeletal muscle (left), the cardiac muscle of the heart (right), and the smooth muscle (center) that is a major component of the walls of blood vessels, the urinary and gastrointestinal tracts, and several other parts of the body.

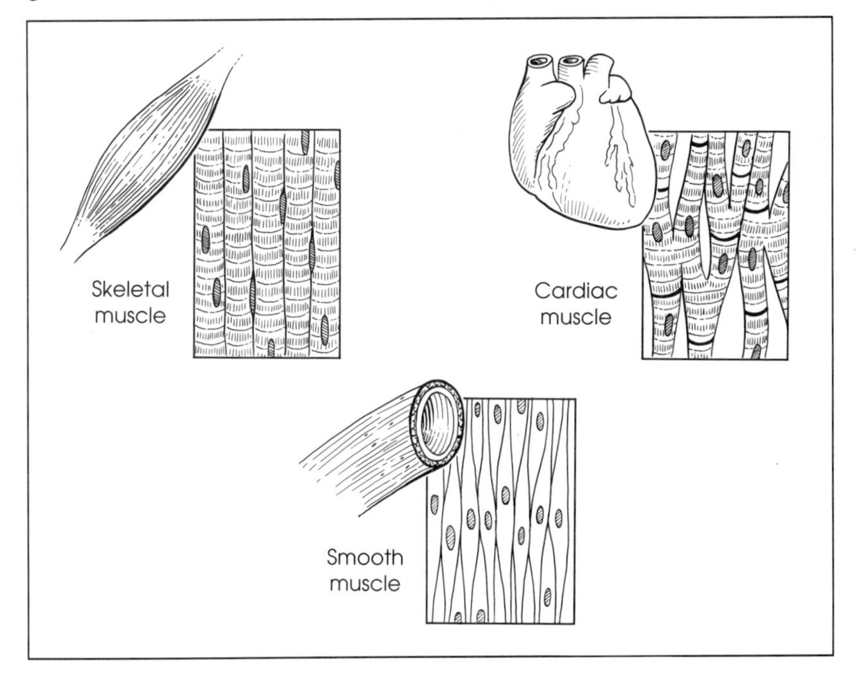

Smooth muscle tissues that do exhibit the domino effect show either of two types of contractions. *Tonic contractions* keep the muscle at least partially contracted at all times. This prevents organs such as the stomach and urinary bladder from becoming completely loose, or flaccid, and distorted. *Rhythmic contractions*, on the other hand, are the pulsed contractions that travel through smooth muscle, kneading and propelling food along the digestive tract.

CARDIAC MUSCLE

Cardiac muscle is found only in the heart, where it is responsible for the contractions that keep the heart pumping at a rate of about 100,000 beats per day, throughout life. Their abundant mitochondria and large supply of oxygen from a great many blood vessels help cardiac muscle cells to keep up this pace, enabling the heart muscle to produce enough ATP to function continuously without acquiring an oxygen debt.

Unlike other muscle cells, cardiac muscle cells are not called fibers. Each is a discrete entity with only one cell nucleus, or at most, two nuclei. Instead, the term *cardiac muscle fiber* refers to a chain of cells linked within the tissue. Like smooth muscle, cardiac muscle is involuntary, but resembles skeletal muscle in having essentially the same contractile process, and its cells have striations resembling those of skeletal muscle fibers. All of the cardiac muscle cells within the heart are interconnected by structures called intercalated disks, which facilitate the rapid passage of contractile impulses from one cell to its neighbor. Consequently, the stimulation of even one cardiac cell by a nerve impulse will cause the entire heart to contract. The result is a network of cells that beat in a synchronous rhythm. A collection of such cooperating cells is called a syncytium.

However, cardiac muscle cells can contract even without nerve stimulation. They do this because between contractions the membrane of the cell accumulates its own electrical charge, which then acts to stimulate neighboring cells. When the heart functions in this way, however, the heart rate is only about half of what it is when the muscle is receiving nerve signals.

CHAPTER 5

SKELETAL MUSCLE AND EXERCISE

A world-champion high jumper demonstrates the strength and coordination of which the skeletal muscles are capable.

Skeletal muscles exist under a constant, low degree of tension. This produces a firmness called tonus, or muscle tone. Although muscle tone is an involuntary feature because the body tenses the muscles automatically, the muscles still require constant use in order to keep a proper amount of such tone. Some of this tone can be maintained through everyday activities, such as walking, bending, or even sitting up in a chair. But more vigorous exercise is needed to produce a high degree of muscle tone, as well as good strength and endurance.

EFFECT OF MUSCLE CONTRACTIONS

When it contracts, a muscle changes shape, usually becoming shorter and thicker. For example, the bicep muscle in the upper arm, which helps to raise the lower arm at the elbow, becomes rounded when it contracts. In some cases, however, a muscle grows longer as it contracts. Bending, for example, tightens yet stretches muscles in the lower and middle back.

One of the mechanisms by which skeletal muscle contraction produces movement. Contraction of the biceps muscle in the upper arm pulls the forearm upward at the elbow. Contraction of the quadriceps muscle in the upper leg pulls the leg upward at the knee.

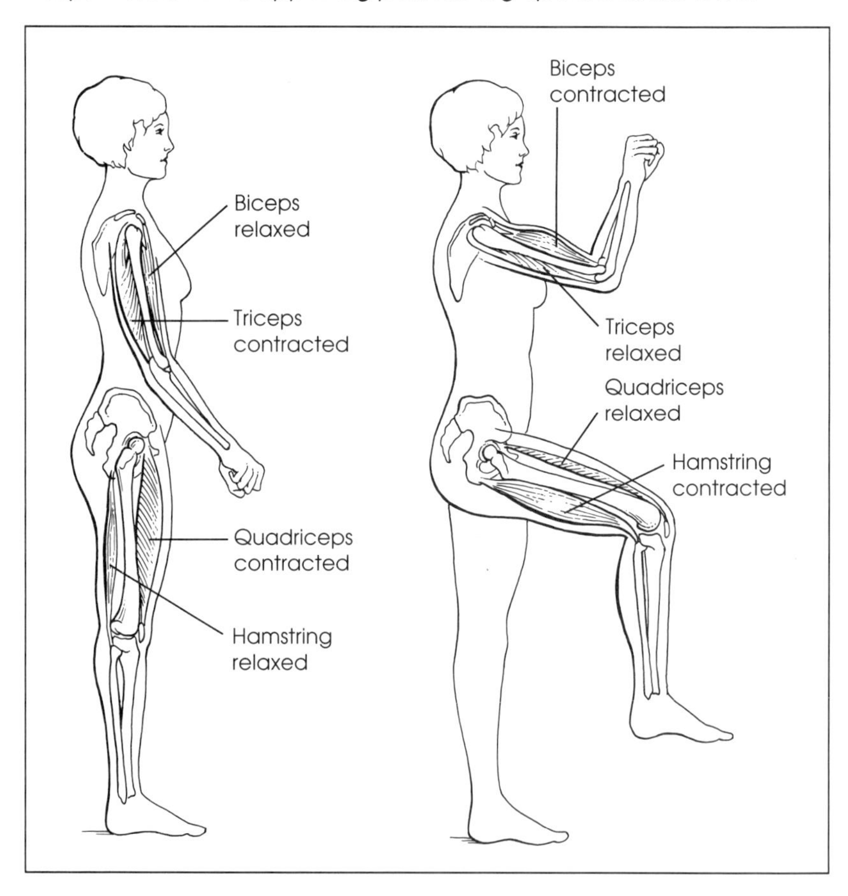

TYPES OF MUSCLE CONTRACTION

Muscles can undergo two different types of contraction: *isotonic* and *isometric*. Isotonic contractions are those that cause movement. Lifting a weight, pushing a wheelbarrow, and picking up a pencil are all activities resulting from isotonic muscle contractions. Isometric contractions do not produce movement. Someone who pushes hard against a brick wall is likely to find that the wall has not moved at all. In this case, the muscles pushing against the wall have undergone isometric contractions.

STRENGTH AND EXERCISE

To a great extent, heredity and gender govern the size of an individual's muscles. Additionally, the male hormone *testosterone* contributes to muscle growth, which, under normal circumstances, is why men have larger muscles than do women.

As is well known, however, exercise plays an important role in increasing muscle size and strength (the amount of force that a muscle can apply in a single contraction).

Increase in Size and Strength

The repeated muscle tension that occurs during exercise forces muscle fibers to form new actin and myosin filaments. As a result, the diameter of each muscle cell increases, in turn causing the entire muscle to grow bigger. This increase in muscle size is called *hypertrophy*. The greater the diameter of a muscle fiber, the stronger its contractions will be. Therefore, larger muscles are stronger muscles. It is also believed that with frequent exercise, some of the largest muscle fibers split into two new cells, so that there is an increase—albeit a small one—in the number of muscle fibers. According to the well-known physiologist Dr. Arthur C. Guyton, proper training can increase muscle mass by 30% to 60%.

However, exercise also builds muscle strength in other ways. For example, it conditions the nervous system to simultaneously stimulate

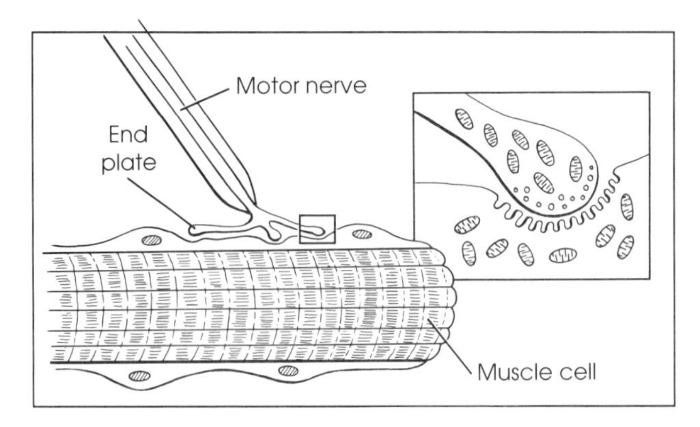

Muscle contraction occurs when the motor nerve connected to a muscle delivers an impulse that causes filaments within the muscle fiber to slide over one another, making the muscle contract.

a greater number of motor units within a particular muscle or set of muscles. Exercise also produces a leaner body, reducing the encumbrance of muscles by fat.

Conversely, if exercise stops, the actin and myosin filaments in a muscle will shrink, the quantity of connective tissue decreases, and the number of blood vessels and mitochondria are reduced. The entire muscle will shrink, or *atrophy*. Prior to old age, the atrophy of muscle tissue does not normally involve a decrease in the number of fibers, and the effects can be reversed. In later years, however, some irreversible atrophy, with a reduction in muscle fiber content, can occur.

EXERCISE AND MUSCLE ENDURANCE

Muscular endurance normally refers to the ability of a muscle to contract repeatedly. The difference between strength and endurance can be understood in terms of weight lifting. A person capable of lifting a 150-pound weight is stronger than someone who can lift a maximum of 50 pounds. But an individual who can lift a 50-pound weight 30 times in a row has more endurance than someone who can lift it only 10 times.

Exercise improves endurance in several ways. These include:

- Increasing the supply of muscle enzymes that allow the muscles to produce more ATP.

- Increasing the number of the small blood vessels known as capillaries, thus allowing fast-twitch muscle fibers to receive a larger blood supply, improving their aerobic capacity and allowing them to operate longer without fatigue.
- Increasing the number and size of mitochondria in the fast-twitch fibers, again benefiting their aerobic capacity.
- Increasing the quantity of myoglobin in muscle cells, permitting them to hold more oxygen.
- Enlarging slow-twitch fibers.

As a result of these changes, even a person whose leg muscles contain a majority of fast-twitch fibers can train to run long distances, while a person with more slow-twitch fibers can become a faster runner.

FORMS OF MUSCLE DEVELOPMENT

The nature of an individual's exercise affects the way that person's muscles develop. Someone who wants to build large muscles has a better chance of doing so by performing exercises that require the

Olympic gold medalist Kristi Yamaguchi is widely regarded by fans of figure skating as the sport's most poised and graceful skater. Her artistry reveals what cannot be quantified by anatomy alone—the sheer beauty of the human form in motion.

muscles to exert maximum force, even if they contract only a few times. In order to improve the shape and definition of the muscles, however, endurance exercises are best. These involve contracting a muscle a great many times, even if the force exerted with each contraction is relatively small.

EXERCISE AND THE ELDERLY

As discussed earlier in this chapter, muscle mass and strength tend to decline with age. Yet research done at the U.S. Department of Agriculture Human Nutrition Research Center at Tufts University indicates that even frail persons in their nineties can improve their muscle size and strength by exercising. This research involved 10 nursing home residents between the ages of 86 and 96. The elderly persons performed leg exercises for eight weeks, with the patients who finished training showing an average 174% improvement in muscle strength. At least five of the subjects also showed larger mid-thigh muscles. The research indicates that loss of muscle strength in the elderly is related to lack of exercise as well as the biological changes that accompany aging.

CHAPTER 6

SKELETAL INJURIES

Without a healthy and properly functioning nervous system, the musculoskeletal system cannot do what it is supposed to do. Former New England Patriots football star Darryl Stingley is now confined to a wheelchair because his nervous system can no longer send proper signals to his muscles.

When a bone is subjected to twisting or compression at a force that exceeds its strength, the bone can break, or fracture. Fortunately, medical science has developed a number of effective treatments for such serious bone injuries.

FRACTURES

Types of Bone Fractures

The nature of a bone fracture depends on the type of object causing the injury and the direction and magnitude of the force to which the bone is subjected.

- A simple fracture occurs when a bone breaks without its broken ends protruding through the skin.
- A compound fracture is one in which the bone breaks through the skin and becomes exposed. This type of fracture is particularly dangerous, since microorganisms can invade the body through the torn skin and cause an infection.
- A transverse fracture runs completely across the bone shaft at a right angle to the length.
- A greenstick fracture does not extend completely across the bone shaft, but instead leaves the bone only partially broken.
- A fissured fracture occurs along the length of the bone shaft rather than running across it.
- An oblique fracture runs across the bone shaft, but at a diagonal angle to the shaft.
- A comminuted fracture is one in which the bone shaft is broken into several fragments.
- An impacted fracture occurs when the broken ends of the bone shaft are pushed into each other.
- A spiral fracture results from the twisting of the bone. The fracture line runs in a spiral pattern around the shaft or axis of the long bone.
- A depressed fracture occurs when the broken portion of the bone is driven inward. This type of injury tends to occur when the flat bones at the top of the skull are broken.

How Bone Fractures Heal

When a bone fractures, the body repairs the damage through a complex series of steps.

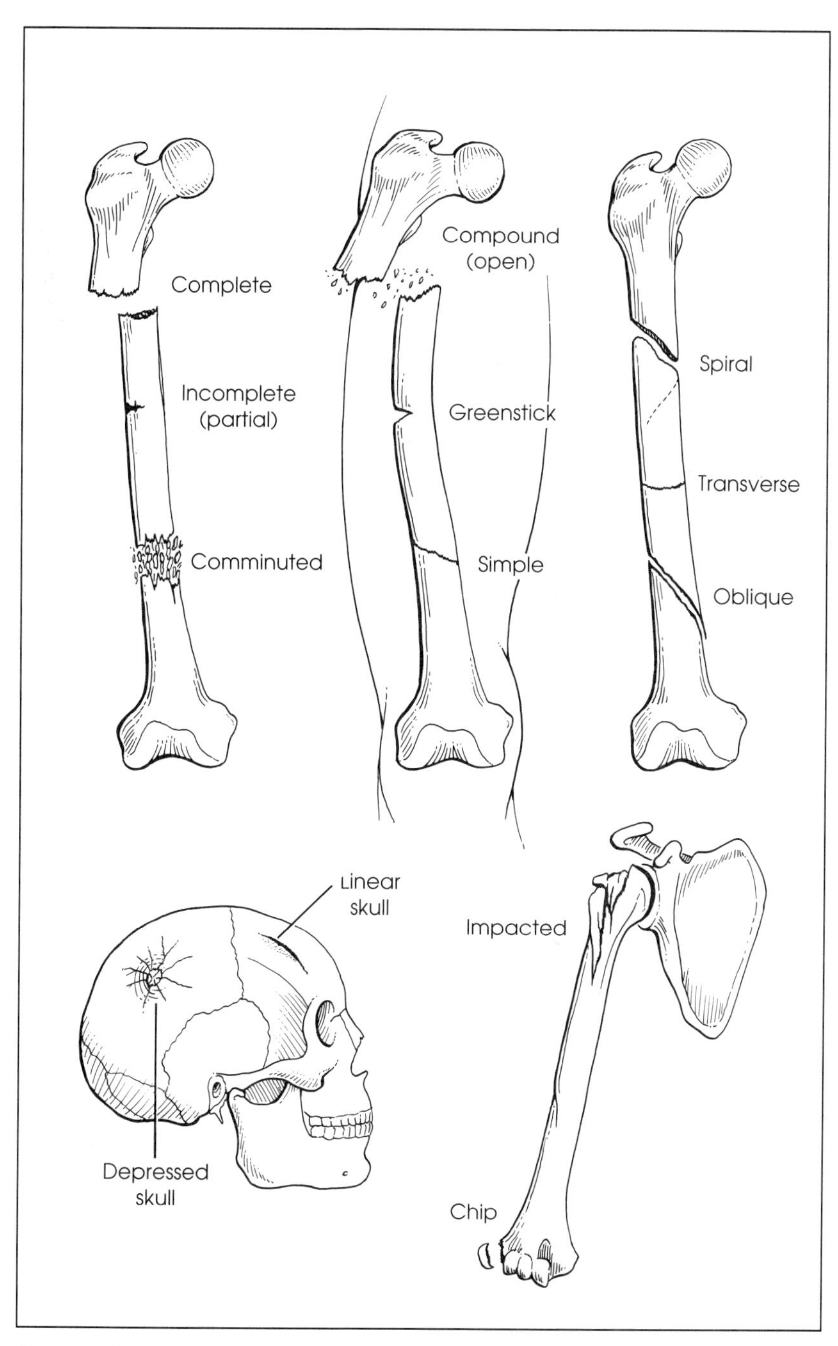

Types of bone fractures.

- When the break occurs, it also ruptures blood vessels within the bone and periosteum. Blood leaks out of these ruptured vessels and forms a blood clot at the injury site.
- As days and weeks pass, new blood vessels begin to extend into the fracture area, and osteoblasts travel from the periosteum to the site of the fracture.
- The osteoblasts multiply, producing new spongy bone. Granulation tissue—a soft, fleshy material composed of a combination of blood vessels and connective tissue—also forms. Additionally, fiber-forming cells called fibroblasts produce fibrocartilage at the fracture site. This material eventually bridges the spaces between the pieces of broken bone, forming a fibrocartilaginous callus.
- Specialized white blood cells called phagocytes devour and digest both the blood clot and any dead or injured cells at the fracture site, while osteoclasts resorb loose bone fragments.
- Eventually, the callus begins to disintegrate, while osteoblasts from the inner layer of the periosteum and from the endosteum move into the spaces that open up during this disintegration. The osteoblasts produce a bony callus to replace the fibrocartilaginous one. New blood vessels also invade the site.
- Initially composed of spongy bone, the bony callus eventually undergoes transformation into compact bone.
- Osteoclasts resorb any excess bone from the repair site.

The time required for a fracture to heal can vary. Normally, the repair takes longer if there is a large gap between the broken ends of the bone. The healing time can also depend on the patient's age. It can take as little as two weeks for a newborn baby's clavicle, or collarbone, to heal, while in a four-year-old the same bone can take three weeks to heal, in a teenager, six weeks, and in an adult, eight weeks.

The bones of older persons take longer to heal because of structural and chemical changes that occur in osteogenic cells as an individual ages. As a result, these cells, while retaining their ability to produce new bone, become less efficient at this task.

Problems with Healing

Fractures do not always heal properly. Delayed healing is said to occur when a bone takes longer to repair itself than expected. This does not mean that the bone will not heal well, but it can indicate that the broken bone is not properly immobilized, that the fracture site is not receiving an adequate blood supply, or that an infection is interfering with healing.

Failure of the fractured ends of a bone to grow back together is called a *nonunion*. One form of nonunion, known as a pseudoarthrosis, occurs when the broken ends of a bone, instead of uniting and becoming rigid, form a flexible jointlike structure.

A *malunion* occurs when a fracture heals in an abnormal position or in poor alignment. In a *cross union*, two parallel bones, such as the radius and ulna of the forearm, become fused together as they heal.

Treatment of Fractures

If the broken ends of a bone have shifted, they must be realigned so that the bone can heal in its correct position. This realignment is called *reduction*. In some cases, an orthopedic surgeon can do this under local anesthesia, using X-ray films and a knowledge of the proper locations of bones and other structures in the injured limb or part of the body. In this case, the anesthesia is used to numb only the site of the injury; the bone does not have to be exposed. This is known as a closed reduction. Sometimes, however, the bone must be realigned surgically—in an open reduction that allows the physician to see the fracture directly.

Two primary methods for immobilizing a fracture are used to hold the broken ends of a bone in their places until they can grow together: *external fixation* and *internal fixation*. In external fixation, the bone is immobilized by attaching a rigid supporting device to the outside of the body at the point where the break exists. This support can take the form of a splint or a plaster of paris cast.

To set a broken limb in a plaster cast, the physician first wraps the limb in cotton padding. The doctor then soaks plaster bandages in water

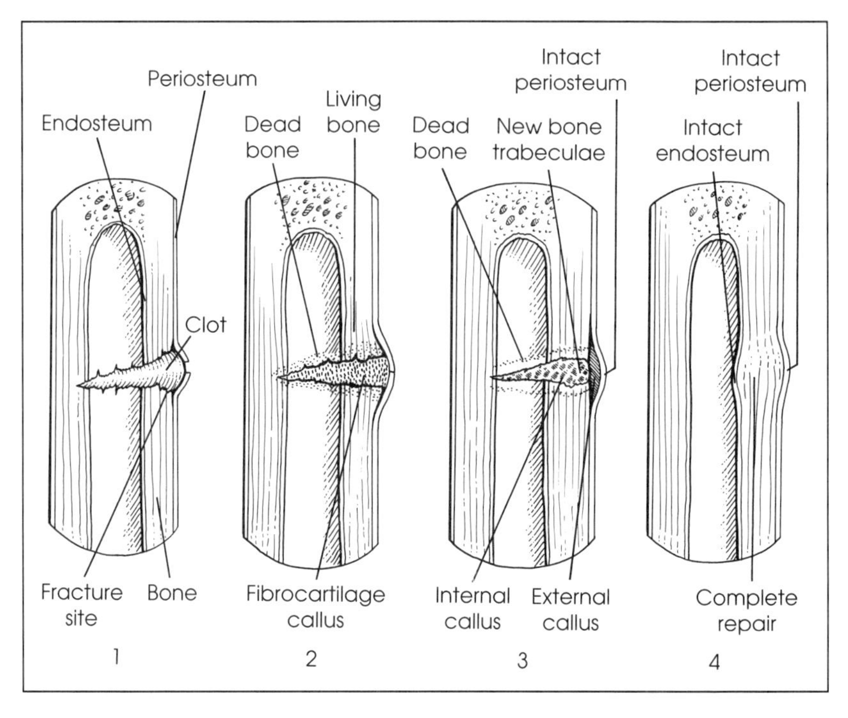

Steps in the healing process of a broken bone. In the first step, fibroblasts produce cartilage that fills the gap between the ends of the broken bone. During and after this step, white blood cells patrol the area of the break, digesting clotted blood and dead or injured cells, while osteoclasts resorb fragments of bone. Next, the cartilaginous material disintegrates and is replaced with spongy bone. In the final step, compact bone replaces the spongy bone.

and wraps them around the padding. Other materials can also be used in casts. For example, a cast can be made from a plastic material known as polyurethane instead of plaster of paris, or it can be constructed from lightweight fiberglass.

Splints also come in several varieties. Although a splint, like a cast, serves as a rigid support, it does not encircle the limb containing a fractured bone, but is instead applied to only one side of the limb. Splints can be constructed from a variety of materials, including plaster of paris, wood, metal, or plastic, and are typically held in place with a bandage.

With internal fixation, a fractured bone is held in place by a metal plate or pin attached directly to the bone. Fractures of the long bones, for example, can be immobilized with an intramedullary nail, which is inserted into the medullary cavity to stabilize a fracture in the middle of the shaft. Internal fixation may be used when a cast or splint will not hold a bone with sufficient rigidity until it heals. A physician may also opt for internal fixation if external fixation poses any sort of danger, such as the risk of pressure sores from a plaster cast.

Healing with Electrical Stimulation

Although surgery may be needed when a broken bone fails to unite, some fractures can be encouraged to heal by stimulating the affected bone with electricity. In such cases, electrodes can be inserted through the skin and applied directly to the fracture, after which a small and steady electrical current is passed through the bone.

Although the precise manner in which electricity brings about the healing of bone is unknown, it may be related to the ability of bones themselves to produce an electrical current under the influence of physical pressure—a property known as piezoelectricity. It is suspected that when a bone is subjected to a bending force, fluid in the bone is squeezed through various canals. The walls of these canals are in turn thought to pull the electrically charged particles of various elements—which are known as ions and carry either a positive or negative electrical charge—out of the fluid within the canals. If negatively charged ions are removed, the remaining fluid will be left with a positive charge, and vice versa. Experiments suggest that this electrical charge stimulates the production of new bone. This in turn strengthens the bone against the forces being applied. This same principle may underlie the effect of externally applied electricity on a fractured bone.

DISLOCATIONS

When the ends of the bone in a joint fail to make proper contact with one another, the joint is said to be *dislocated.* This can result from a direct blow or other force that impacts on the joint, from a congenital defect (a defect present at birth), or from a disorder that develops later in life, such as an infection or *rheumatoid arthritis.* A joint may even become dislocated as the result of an unnatural force applied by the muscles surrounding the joint. Thus, the temporomandibular joint, which hinges the jaw to the skull, can become dislocated as a result of a wide yawn or unrestrained laughter. The partial dislocation of a bone from its normal position in a joint is known as a subluxation.

The bones in a dislocation must be guided back into place in order to restore proper function to the affected joint and prevent muscle spasm and harm to nerves, tissues, and blood vessels within and around the joint. This procedure must be performed by trained medical personnel. Once the joint has been relocated, it may have to be immobilized for two to eight weeks.

Despite this, new evidence suggests that immobilization may not be the ideal way to treat a dislocated joint. It has been found that the injury from a dislocation may heal more quickly if the joint is kept in constant motion. This maintains the flow of synovial fluid, which lubricates the joint and prevents the surrounding synovial membrane from becoming attached to the cartilage that cushions the joint. A motorized, continuous passive-motion device can be used to keep the joint moving and prevent the patient's muscles from becoming fatigued.

SPRAINS

Technically, the injury known as a *sprain* is not an injury to the bone but to the ligaments connecting two bones at a joint. In such an injury, the ligaments are stretched or torn when the joint is twisted. A mild sprain can normally be treated with rest and ice, and later with heat, as well as by bandaging the joint to immobilize it and support it. A sprain may heal in a few days, although weeks or months may be needed for healing if ligaments have been torn, and in some such cases surgery may be required.

CHAPTER 7

SKELETAL AND JOINT DISORDERS

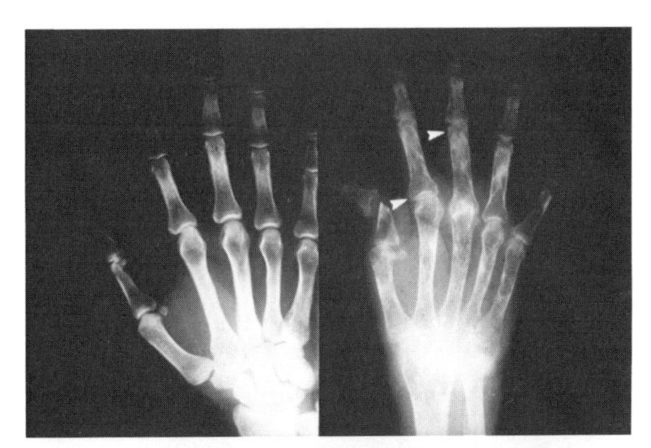

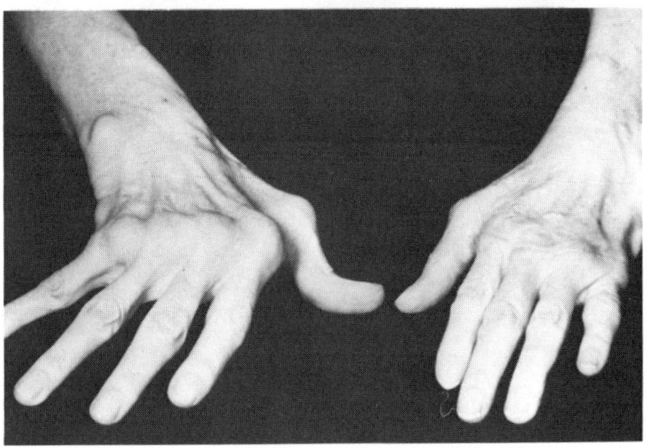

X-ray images of a normal hand (top)and a hand afflicted with rheumatoid arthritis (bottom).

Despite the great advances made since 1900 in the understanding of bone and joint disease, physicians still have much to learn concerning the cause, prevention, and cure of such disorders. Nevertheless, current medical understanding can help prevent the onset of crippling ailments over the long term. This chapter will examine some of the more common bone and joint diseases.

OSTEOPOROSIS

Osteoporosis is a common problem among elderly persons, particularly women, in whom it is five times more common than in men, especially affecting women over the age of 65. The disease apparently occurs when the formation of new bone begins to slow down while the body continues to resorb old bone at its previous rate. This leaves the bones porous and easily breakable.

Because relatively few men suffer from osteoporosis, it may be related to menopause. As mentioned earlier, when menstruation ceases after menopause, the secretion of estrogen in a woman's body decreases, allowing parathyroid hormone to release greater amounts of calcium from the bones. Chapter 6 examined other reasons why women tend to lose more bone than men do as they age, and these factors also appear to contribute to the greater vulnerability of women to osteoporosis. Other factors contributing to bone loss are a high phosphorus content in the diet, which was also previously discussed, and alcohol abuse, since alcohol inhibits the body's ability to absorb calcium from the diet.

Race, and apparently heredity, play roles in the onset of osteoporosis. The disease is especially common among whites and Asians, and less common in blacks. As discussed in chapter 3, blacks tend to have a particularly high mineral content in their bones and appear to have a lower rate of bone turnover, with both the formation of new bone and the resorption of old bone occurring more slowly than in whites or Asians. Research suggests that this low rate of turnover may reduce the risk of osteoporosis.

Symptoms and Consequences

Although osteoporotic bones are fragile and fracture easily, most patients feel no symptoms of the disease until a bone actually breaks. The thoracic and lumbar portions of the spine, behind the chest and stomach and in the lower back, respectively, are common fracture sites because of the stress they must bear. The lower spine, for example, consists of a relatively small amount of bone that must support the

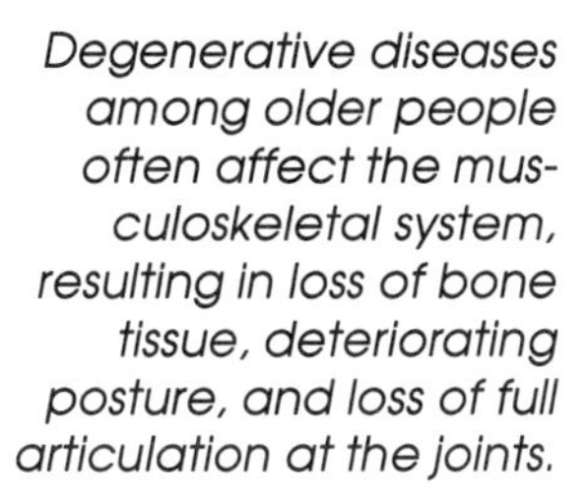

Degenerative diseases among older people often affect the musculoskeletal system, resulting in loss of bone tissue, deteriorating posture, and loss of full articulation at the joints.

entire upper body, while the middle of the spine, in the thorax, is subjected to constant bending, such as leaning forward to read or write.

As bones affected by osteoporosis become increasingly fragile, even a relatively simple movement such as turning in bed can cause a tiny break in one of the lower spinal vertebrae. The initial symptoms of such a fracture may be no more severe than a constant backache. However, repeated fractures of the vertebrae can cause the spinal column to shrink, so that the patient literally becomes shorter. This may also lead to the condition known as kyphosis, in which the thoracic portion of the spine develops an abnormally severe outward curvature.

Another common site of bone fracture in osteoporosis is the neck of the femur—the short section of the upper bone of the leg that leads from its main shaft to the point at which the bone forms a ball-and-socket joint with the hip.

Prevention

Although osteoporosis can be treated, it is preferable and more effective to prevent it altogether. One factor that plays a valuable role in preventing the onset of osteoporosis is regular exercise. Not only can such exercise increase bone production, it can also slow the loss of old bone as an individual ages.

Consuming adequate amounts of dietary calcium is also essential to preventing osteoporosis. Although the U.S. Recommended Dietary Allowance of calcium for women aged 25 and older is 800 milligrams daily, many experts suggest that prior to menopause, a woman should consume 1,000 milligrams per day, and as much as 1,500 milligrams daily after menopause. The administration of estrogen after menopause is also thought to help prevent excessive bone loss. However, because estrogen has been linked to serious side effects, including cancer, physicians tend to administer the hormone in low doses. The hormone calcitonin, normally secreted by the thyroid gland but also prepared and taken as a medication, is another treatment used to prevent bone loss. It counteracts the calcium-resorbing effect of parathyroid hormone on the bones and, unlike estrogen, can be useful in men as well as in women. It may also be a preferable alternative to estrogen therapy for women who are especially susceptible to the undesirable side effects of such therapy.

Among young persons, reducing the consumption of soft drinks, snack foods, and other high-phosphorus items can help build strong bones and withstand later bone loss. Because alcohol can aggravate bone loss, it also follows that avoiding it or consuming it in moderation can help maintain skeletal strength.

ARTHRITIS

The Arthritis Foundation estimates that 37 million Americans, including almost 50% of people aged 65 or older, suffer from *arthritis*, an inflammation of the bone, cartilage, and other tissues of a joint, with potentially crippling consequences. This section will examine two common forms of arthritis: osteoarthritis and rheumatoid arthritis.

Osteoarthritis

The most common form of arthritis, osteoarthritis is most often caused by the repeated stress of using the body's joints over the course of a lifetime. According to the U.S. National Institutes of Health (NIH),

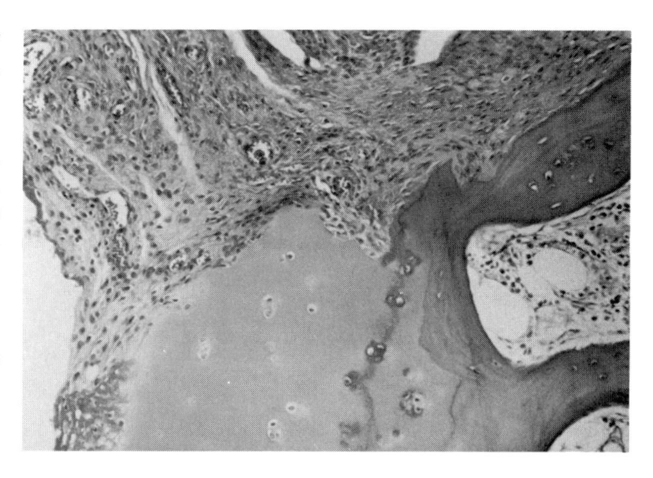

Microscopic view of the pannus that forms within diseased joints in rheumatoid arthritis. Composed of connective tissue and the synovial membrane of the joint, the pannus may eventually fill the space between the bones of the joint, fusing the bones and preventing them from moving.

an estimated 15.8 million Americans suffer from osteoarthritis, also known as degenerative joint disease. Weight-bearing joints, such as the knees and hips, are particularly vulnerable to the disease, with the spine and fingers also common targets. Although the symptoms of osetoarthritis usually do not begin until age 60, the joint deterioration that characterizes the disease probably begins in a person's late twenties.

Although osteoarthritis may ordinarily result from mechanical stress on the joints, researchers believe that some individuals have a greater tendency than others toward developing this disease. It is thought that in such cases, enzymes bring about the deterioration of cartilage in the joint, leading to its replacement with a less resilient cartilage that offers inferior cushioning and protection. It has been suggested that this breakdown-replacement process may occur more readily in some persons, causing their joints to deteriorate at a faster rate.

Heredity may also have a more obvious role in osteoarthritis. One Ohio family was found to have a genetic mutation that affected the production of collagen II, a cartilage-strengthening protein. Family members who inherited this mutation began to experience pain and stiffness in their joints while still in their teens and twenties.

Other factors contributing to osteoarthritis include injury and infection of one or more joints; skeletal misalignment that puts abnormally high stress on the cartilage of a joint (some individuals are born with this condition); and excess weight, with obese persons at greater risk of developing osteoarthritis because of the greater gravitational forces on their joints. A long-term study of people in Framingham, Massachusetts, found a strong relationship between obesity and the development of osteoarthritis of the knee among women.

Osteoarthritis is often treated with a mild *analgesic*, or pain-relieving, drug such as aspirin, although in severe cases, drugs known as corticosteroids, which are synthetic relatives of the body's own steroid hormones and which combat inflammation, may be prescribed. Exercise to strengthen muscles around the joint and keep the joint as mobile as possible is also important, and heat treatments or cold compresses can help to ease the pain of osteoarthritis. When such treatments fail, surgery may be needed to remove areas of bone and cartilage that are causing inflammation within a diseased joint. In some cases it is necessary to replace the entire joint with an artificial joint made of metal and plastic.

The X ray on the left shows a knee joint in an advanced stage of arthritis. At right, the diseased bone has been replaced by a prosthetic joint made of metal and plastic, restoring the patient's mobility and dramatically reducing the level of pain.

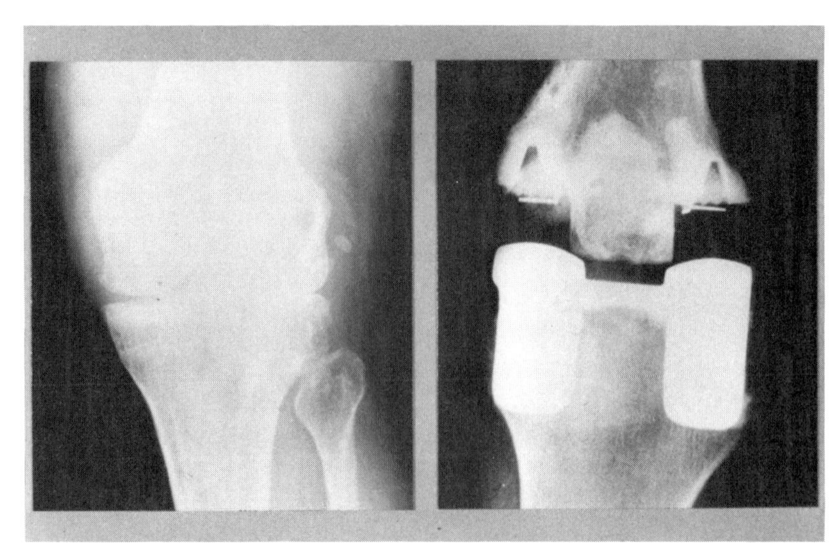

Rheumatoid Arthritis

Rheumatoid arthritis can have a far wider impact on the body than osteoarthritis. It affects other tissues and organs of the body as well as the joints, and in combination with joint pain it can cause severe fatigue, weakness, weight loss, and fever. The NIH estimates that about 2.1 million Americans suffer from rheumatoid arthritis. Moreover, approximately half of all patients affected by the disease become disabled in some way, and 10% to 15% fall victim to a particularly severe form of the disease. People between the ages of 20 and 50 are common targets of rheumatoid arthritis, although no age group entirely escapes the condition. Although it is not known why, 75% of those affected by rheumatoid arthritis are women.

Rheumatoid arthritis is an autoimmune disease. This means that the immune system, which normally fights invading bacteria and viruses, actually turns against the body itself, attacking healthy tissue. In rheumatoid arthritis, the synovial membrane of a joint is the first site of attack and eventually thickens and develops folds. The attack on the tissues of the joint also leads to the production of substances that dissolve bone and cartilage. In an effort to repair the damage, a structure known as a pannus, composed of connective tissue and the thickened synovial membrane, forms within the joint and may eventually fill the joint space, fusing the bones of the joint to one another.

Outside the joints, the autoimmune reaction of rheumatoid arthritis can create an inflammation of blood vessels, the skin, nerves, muscles, and the heart and lungs, although such wide-ranging consequences are unusual.

Some scientists suggest that rheumatoid arthritis begins when a virus or some other infectious agent enters the body and elicits an attack not only on itself but also on the body's own cells by disease-fighting cells of the immune system. Heredity may contribute to this, causing some individuals to be particularly susceptible to rheumatoid arthritis. Researchers also suggest that exposure to certain minerals or chemical substances may contribute to the development of rheumatoid arthritis. Long-term studies have, for example, found a high incidence of rheumatoid and certain other types of arthritis among coal miners and laborers who have worked with granite or asbestos. One study also indicates a connection between smoking and rheumatoid arthritis.

The treatment of rheumatoid arthritis is directed at controlling the inflammation of the joints and minimizing their destruction. Along with relatively minor pain relievers such as aspirin, physicians also prescribe drugs such as methotrexate, originally developed for the suppression of cancer cells. These drugs are capable of regulating the immune response by suppressing the immune system. In suppressing immunity, however, these medications leave the patient vulnerable to infection. Researchers hope to find drugs capable of regulating only certain components of the immune system while leaving others unaffected, so that patients with rheumatoid arthritis are also better able to fight other diseases.

In addition to drugs that suppress the immune system or parts of it, corticosteroids are also used to combat the inflammation in rheumatoid arthritis, although they too can pose dangers. If administered over a long period of time, for example, corticosteroids can cause loss of bone

This research worker prepares to use an electron microscope in a project designed to study how cells release chemical substances that may promote arthritis.

tissue and weakening of the bones, as well as cataracts, diabetes, and increased vulnerability to infection. The precious metal gold is another medication for rheumatoid arthritis. Although researchers are not fully certain as to how it works, gold reduces both the inflammation and pain in many cases of the disease.

As with osteoarthritis, treatment that does not involve drugs also plays an important role in treating rheumatoid arthritis. A combination of rest and exercise can help to keep joints flexible and strengthen surrounding muscles.

Although current treatments, including those that use drugs and those that do not, can aid about 85% of persons with rheumatoid arthritis, they may not help the most severely affected individuals. These patients, like some of those with osteoarthritis, may have to have a diseased joint surgically replaced with an artificial one.

SPINA BIFIDA

Spina bifida is a spinal deformity that results when the rear portions of the vertebrae do not completely close around the spinal cord. One of a group of conditions called neural tube defects, in which the brain and spinal cord do not develop normally, spina bifida can result in an inability to control the bowels and bladder, difficulty with movement, and paralysis. In the United States, 1 out of every 1,000 infants born has a neural tube defect, according to the U.S. Centers for Disease Control. Although the exact cause of spina bifida is unknown, genetic factors, poor diet, and even geographic location may all play a role in its occurrence. Moreover, valproic acid, a drug used to prevent epileptic convulsions, has been linked to an increased incidence of spina bifida in the infants of women who take this drug.

A common blood test called the maternal serum alpha-fetoprotein screening test, which is performed during the 16th to 18th week of pregnancy, can provide some indication that a fetus is suffering from spina bifida. This is because neural tube defects permit large amounts of alpha-fetoprotein, a protein produced by the fetus, to escape into the mother's bloodstream. However, the test is not definitive proof of a defect. Further testing is required to make such a diagnosis.

The strain that an infant undergoes during delivery may worsen the paralysis associated with spina bifida. A study reported in 1991 from the Swedish Hospital Medical Center in Seattle, Washington, provides evidence that delivering such infants via cesarean section, in which the fetus is removed surgically through the abdomen, can reduce the chance of severe paralysis. The defect in the infant's spine can then be repaired by surgery immediately after delivery, in order to limit complications related to spina bifida.

BONE CANCER

Bone cancer consists of tumors that destroy healthy bone tissue, weakening the skeleton and causing the bone to fracture. In some cases, the cancer originates in another part of the body, such as the prostate gland, breast, or lung, and travels, or metastasizes, to the skeleton by way of the bloodstream. In other instances, the cancer originates in the bone tissue itself. In such instances the disease occurs most often in the legs, particularly in or around the knees. Conversely, bone cancer may spread from the bone to other organs before it is detected.

Bone cancer most often occurs in adolescents, during the years in which they undergo their greatest growth spurts. This form of bone cancer is known as *osteogenic sarcoma* and its development suggests that sections of bone undergoing rapid growth are most susceptible to this disease. Symptoms of osteogenic sarcoma include pain and swelling in the bones and joints.

Myeloma, a disease more prevalent in women than in men, is the most common type of bone cancer. In this type of cancer, malignant tumors in the bone marrow hinder the generation of red blood cells. Symptoms of this condition include anemia and osteoporosis.

Bone cancer is often treated by removing the diseased section of bone—in some cases by amputating a limb and in others by cutting away the tumor but preserving the limb, after which anticancer drugs are given to kill any cancer cells that may have spread to other parts of the body.

OSTEOMYELITIS

The noncancerous disease known as osteomyelitis is an infection of the periosteum, bone tissue, and marrow of the long bones. It typically begins before adolescence. In the metaphysis it extends along the Haversian canals and marrow cavity, and underneath the periosteum, killing osteocytes and destroying the channels that carry blood vessels through the bone. Although osteomyelitis may be caused by a wide variety of microbial organisms, many cases result from invasion of the bones by *Staphylococcus aureus*, a species of bacteria that ordinarily resides on the skin but can enter the body through a cut or scrape. As the disease spreads along the bone shaft before moving to the epiphyses at the ends of the bone, it destroys bone tissue. It then spreads through the hyaline cartilage and into the joints.

Osteomyelitis can be either *acute* or *chronic*. The severe but short-term, acute form of the disease most often occurs in children during a rapid period of growth. The chronic, long-term form is more common in adults. Although osteomyelitis is potentially fatal, it can be effectively treated with antibiotic drugs.

RICKETS AND OSTEOMALACIA

The terms rickets and osteomalacia refer to the same condition, the difference being that rickets is the childhood form of the disease and osteomalacia the adult form. In both cases, the disease results from a lack of vitamin D, which, as discussed earlier, is essential in helping the body to absorb calcium from the intestines into the bloodstream. Because calcium is essential in building bone, a lack of vitamin D ultimately interferes with formation and renewal of the skeleton, causing bones to soften. In children with rickets, the failure of adequate amounts of calcium to be deposited at the epiphyseal growth plates can prevent the proper lengthening of growing bones, stunting overall growth.

Various factors can create a vitamin D deficiency. In some cases, the diet may not contain enough of the vitamin (which is found in such foods as milk, eggs, and liver). In others, the body may not properly absorb the vitamin. Because ultraviolet light from the sun is essential to the conversion of 7-dehydrocholesterol—a chemical precursor of vitamin D—into the vitamin itself, a lack of sunlight can also lead to a vitamin deficiency. And because heavily pigmented skin absorbs less ultraviolet light than does lighter skin, rickets most often occurs in black children.

The skeletal deformities that result from rickets can include bulging of the front of the skull and weakened, bowed legs, both of which may become permanent malformations. Adults with osteomalacia can also develop skeletal deformities, although in this case large doses of vitamin D can eliminate these abnormalities.

PAGET'S DISEASE

As with osteoporosis, *Paget's disease*—named for the British physician Sir James Paget (1814–99), who first described it in 1876—involves a malfunction of the process through which old bone cells are destroyed and new ones are formed. In Paget's disease, bone is both destroyed and replaced at an excessive rate. This leads to enlargement of the affected portion of the skeleton, with weakening of the diseased bone. The skull, pelvis, vertebrae, thigh bone, and shinbone are the bones most commonly affected. Paget's first encounter with the disease, in 1856, involved a patient complaining of leg pains that had persisted for two years. The patient's legs had gradually become bowed while his height had decreased and his head had grown in size. An autopsy later revealed that the patient had a pockmarked skull that had grown four times thicker than normal.

Although a mild form of Paget's disease may cause no symptoms, severe cases can produce intense pain as the enlarged, diseased bone compresses nearby nerves. Eventually, the legs and spine can become bent, and as the enlarged skull pushes against the auditory nerve, which is essential for carrying sounds from the ears to the brain, the victim's

Sir James Paget, the British physician who first described the disease that bears his name. The rapid destruction and replacement of bone in Paget's disease causes pain, weakening of bone, and skeletal deformity.

hearing may suffer. Some loss of hearing occurs in as many as 50% of persons in whom Paget's disease affects the skull.

The severe, crippling variety of Paget's disease is, fortunately, rare. It tends to occur after age 40, and most often after the age of 60, and is most prevalent among white men. The hormone calcitonin and other drugs that interfere with the bone destruction-replacement cycle in Paget's disease are used to treat the disease. Surgery is used to relieve the compression of nerves and other tissues caused by overgrown bone.

Some physicians believe that the great German musician and composer Ludwig van Beethoven (1770–1827) suffered from Paget's disease, pointing to his short legs and large head, with its asymmetrical shape and overhanging brows. It may in fact have been this illness that caused Beethoven's deafness later in life.

SCOLIOSIS

Scoliosis refers to a side-to-side curvature of the spine that becomes progressively worse. The disorder occurs in the middle, thoracic region of the spine or in the lower lumbar region and affects infants, children, and adolescents. Although infantile scoliosis tends to occur more often in boys than in girls, the adolescent form of the disease is more common in girls. In most instances the cause of scoliosis is unknown, and the disease is said to be *idiopathic*. In a smaller number of cases, genetic abnormalities cause the condition by promoting irregular bone growth in the vertebral bones of the spine. In still other cases, scoliosis may be *congenital* and due to spina bifida and other defects of development.

Without treatment, scoliosis can cause the vertebrae to twist or rotate in relation to one another until the ribs are pushed together on one side of the body and pulled apart on the other. As this happens, the organs within the abdomen and thorax may be pushed out of place or compressed, and as in osteoporosis, the front-to-back curvature of the spine known as kyphosis can also develop, so that the shoulders become rounded, the back becomes hunched, and the chest is sunken inward.

The progressive spinal curvature caused by scoliosis can be stopped by fitting the patient with a spinal brace. If the condition is particularly severe, the vertebrae may be straightened surgically and held in place with rods.

CHAPTER 8

MUSCULAR DISORDERS AND INJURIES

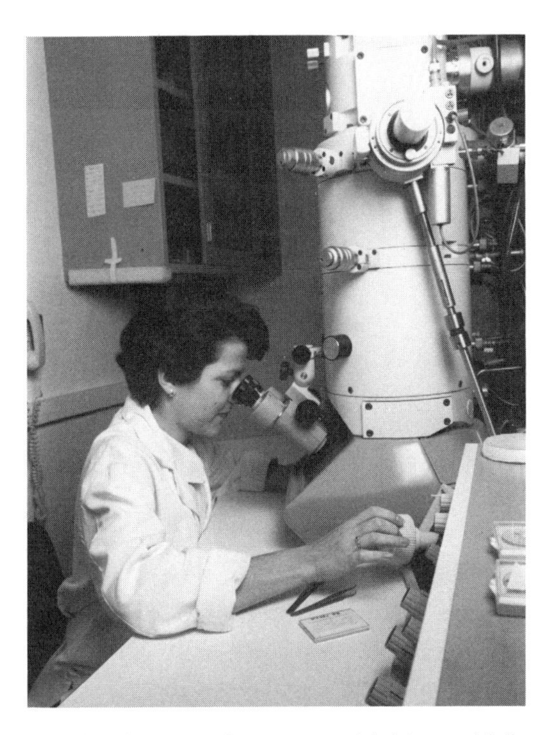

Medical research now uses highly sophisticated devices such as this electron microscope, which led to the discovery of dystrophin, a protein absent in people suffering from muscular dystrophy.

Despite their ability to adapt and grow through conditioning, muscles, like the skeletal system, are prone to injury and disease. Muscle disorders can range from relatively minor problems, such as cramping, to crippling and lethal diseases involving the muscles and their motor neurons.

MINOR DISORDERS

Shin Splints

Among the less serious muscle ailments is the painful condition known as a shin splint. This results from increased pressure in tissues surrounding blood vessels in the region of the shin, compressing these vessels and interfering with the blood supply to the region. The condition is common in dancers, athletes, and other persons who use their feet and legs strenuously and often, and it results from the repeated application of force to the legs during running or jumping on a hard surface. Shin splints are typically treated with rest, by applying ice to the injured area, and by supporting the shins and lower legs with tape or bandages.

Muscle Cramps

Cramps are painful, sudden spasms of a muscle that commonly occur when the muscle tires from exercise. They may also result from keeping a muscle in the same position for too long, as in the case of the well-known writer's cramp in the thumb and first two fingers of the hand, which results from holding a pen or pencil for an extended period. Research has shown that during a cramp, the motor units of a muscle—consisting of the nerve endings on the muscle and the muscle fibers to which these nerves are connected—deliver impulses much more rapidly than in even the most extreme voluntary contractions.

Cramps can be treated by massaging the cramped area, applying heat, or stretching the affected muscle. It is thought that stretching may stop a cramp by inhibiting the transmission of impulses from the motor neuron to the muscle fibers with which the neuron connects. To avoid cramps, it is important to make certain that muscles do not become overtired.

MUSCULAR DYSTROPHY

Among the more serious diseases of muscle is the degenerative disease known as *muscular dystrophy*. Although it falls under a single name, muscular dystrophy actually consists of a group of diseases called *myopathies*, which are inherited and lead to the destruction of muscle and its replacement by fatty tissue. At the cellular level, some of the muscle fibers affected by muscular dystrophy shrink, while others enlarge, possibly as a means of attempting to compensate for the shrunken, weaker cells. The different kinds of muscular dystrophy are classified according to the way in which they are inherited, the age at which they commonly appear, and the rate at which they progress.

Duchenne Muscular Dystrophy

The most severe form of muscular dystrophy is *Duchenne muscular dystrophy*. In order to understand how this disease develops, it is necessary to have a brief discussion of genetics.

Human cells contain 46 *chromosomes*, the long strands of specialized chemical molecules that contain the structural code or blueprint for all of the body's substances. Within the chromosomes are the *genes*—close to 100,000 of them in each human cell. Each gene carries the structural code for a single specific substance, and during normal development, it works in conjunction with other genes to ensure that the individual evolves properly.

These two microphotographs show the difference between normal muscle tissue and dystrophic muscle tissue.

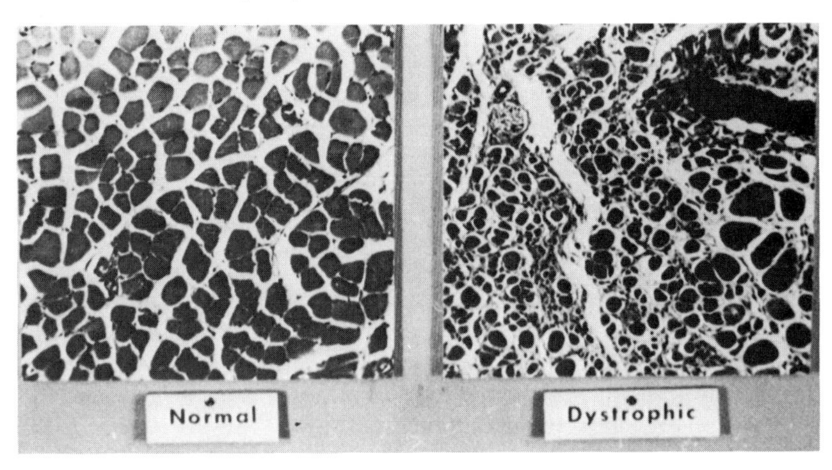

Duchenne muscular dystrophy is caused by a genetic abnormality in the *X* chromosome, one of two different chromosomes, designated *X* and *Y*, respectively, that determine an individual's sex. A child that inherits two X chromosomes, one from its mother and the second from its father, will always be female, while the inheritance of an X chromosome from its mother and a Y chromosome—which is always carried and passed on only by males—from the father will render the child male.

The genetic abnormality that is responsible for Duchenne muscular dystrophy is unlikely to express itself in a female child, because at least one of the child's X chromosomes is likely to be normal and will therefore prevent the dystrophy gene, which is said to be *recessive*, from causing the disease.

By contrast, a male child whose X chromosome carries the gene for Duchenne muscular dystrophy does not have a second X chromosome to counteract it and will develop the full form of the disease.

Such a child will be unable to manufacture dystrophin, a protein found in the sarcolemmal membrane of the muscle fiber that may play a role in maintaining the structure of this membrane. Becker muscular dystrophy, a somewhat less severe form of the disease than Duchenne dystrophy, occurs when the gene for dystrophin manufactures an aberrant form of this essential protein.

The gene for muscular dystrophy was identified in 1986 by Dr. Louis Kunkel and his co-workers at Harvard Medical School and Massachusetts General Hospital. A year later, they also identified dystrophin as the protein product of this gene. Yet precisely how the absence of dystrophin is related to the development of Duchenne muscular dystrophy remains uncertain. One theory is that the absent or missing dystrophin permits an abnormal buildup of calcium within the muscle cell, which in turn sets off the manufacture of enzymes that destroy the cell.

In some cases, Duchenne muscular dystrophy is evident at birth, so that the newborn infant seems "floppy," with poor muscle coordination. In most instances, however, children with Duchenne muscular dystrophy begin to show outward symptoms of the disease—chiefly in the form of difficulty in running and climbing—by the age of three.

Another early sign of Duchenne muscular dystrophy is, paradoxically, an enlargement rather than wasting of the muscles of the calf. This is because these muscles develop an excessive amount of fat and scar tissue and may grow larger to compensate for other muscles that have become weakened by the disease. The deltoid muscles in the shoulder and quadriceps muscles in the thigh can also become enlarged. Despite its effects on these and other muscles, however, Duchenne muscular dystrophy does not affect the muscles of the head.

By age ten, most children with Duchenne muscular dystrophy have experienced severe muscle weakening and damage to their joints and are unable to walk. The loss of muscle tissue proceeds until it affects the muscles of the chest and diaphragm that are essential to breathing. This problem worsens as the intercostal muscles between the ribs are weakened by the disease. Loss of the capacity to breathe may either lead to death from asphyxiation or, more commonly, to respiratory infection and death from pneumonia.

Beyond the Duchenne and Becker forms of muscular dystrophy are several other types of the disease, each with its own characteristics. Facioscapulohumeral dystrophy results from a dominant gene and therefore develops in all persons who inherit this gene. This disease typically begins with a wasting of the facial muscles, leading to a gradual loss of facial expression and then spreading to the muscles of the shoulders and upper arm. Myotonic dystrophy is related to a dominant gene on chromosome 19 and therefore resembles facioscapulohumeral dystrophy in that it affects all persons who inherit this gene.

It is not known whether myotonic dystrophy results directly from the faulty gene, or whether the defective gene is itself the result of another abnormality. However, the disease is characterized by multiple replications of a sequence of three specific nucleotides—the molecules of which genes are constructed. Abbreviated CTG because it consists of the nucleotides *c*ytosine, *t*hymine, and *g*uanine, this three-nucleotide sequence is typically repeated more than 50 times within the gene identified as causing myotonic dystrophy. The symptoms of myotonic dystrophy are similar to those of Duchenne muscular dystrophy and also include cardiac abnormalities. By comparison, it is found far less frequently in the corresponding gene of persons without the disease.

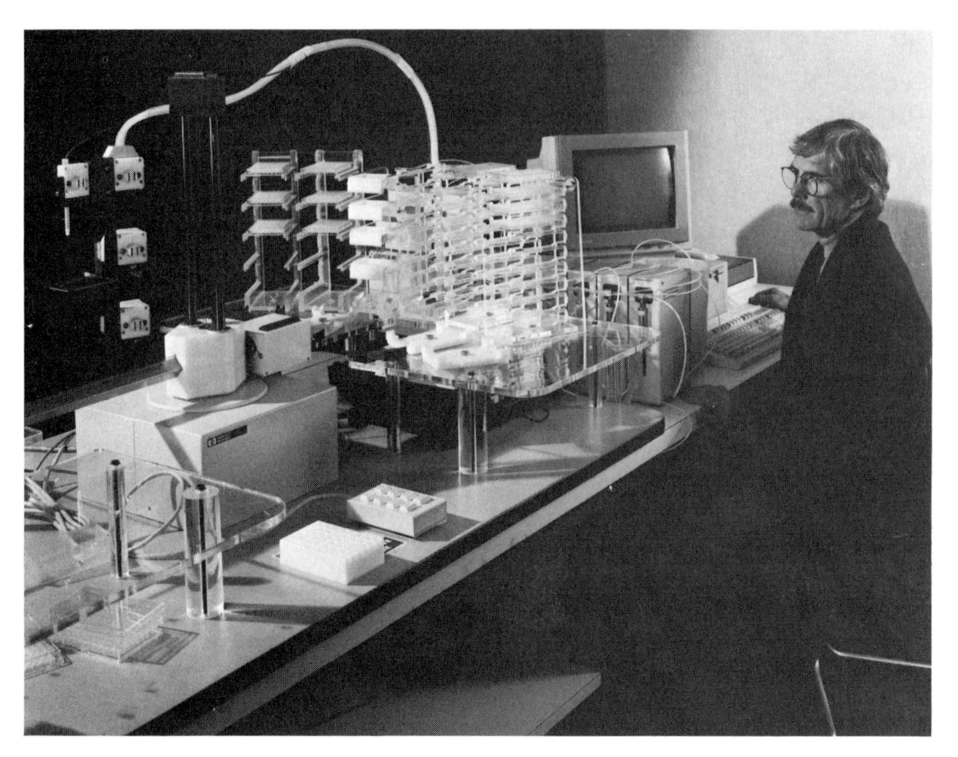

A medical scientist uses a computer-controlled robotic device as part of a project to map the location of genes on human chromosomes. This project may provide invaluable information in predicting the development of genetically determined diseases.

MYASTHENIA GRAVIS

Like rheumatoid arthritis, the disease known as *myasthenia gravis* is an autoimmune disease. It occurs when the substances known as antibodies, ordinarily produced by the immune system in response to alien viruses or bacteria, instead attack the specialized structures known as acetylcholine receptors, which exist on muscle fibers and into which acetylcholine fits like a key into a lock in delivering its messages from the motor nerve to the muscle fibers. Consequently, the immune system cripples or destroys the neuromuscular junction that connects motor nerve endings to the sarcolemmal membrane. Scientists are not certain what triggers the attack, but it interferes with the transmission of nerve impulses to the muscles. In some cases the muscle may simply stop responding to acetylcholine and cease contracting in response to motor-nerve impulses. Another effect of the disease may be an excess of acetylcholinesterase, the enzyme responsible for breaking down acetylcholine after it has finished transmitting its messages. This excess of acetylcholine is the basis for the medical treatment of myasthenia gravis, which employs drugs that inhibit the action of acetylcholinesterase, thereby preserving acetylcholine within the neuromuscular junction and helping to maintain the transmission of nerve impulses to the muscle.

Before the advent of such treatment, as well as immunosuppressive treatment directed at stopping the body's autoimmune attack on its acetylcholine receptors, myasthenia gravis was lethal among more than 60% of its victims. Today it is fatal in only 5% of cases.

Myasthenia gravis is three times more common in women than in men, occurring mainly in women between the ages of 20 and 40. When it does strike men, it tends to appear after age 40. Although it affects various muscles with different degrees of intensity, the disease is characterized by extreme muscle fatigue with even the slightest exertion. For unknown reasons, it often affects muscles in the head, especially those that move the eyes, eyelids, and other eye-related structures, causing drooping of the eyelids and double vision. Research in Japan has recently suggested that the high frequency with which myasthenia gravis attacks the eye muscles may come from the greater numbers of motor-nerve terminals on each of these fibers than are present on the fibers of other muscles, and from structural differences between these receptors and those of other muscle fibers.

When it affects muscles in the face, neck, throat, and larynx (voice box), myasthenia gravis may cause difficulty in speech, chewing, and swallowing. It may also attack the arm and leg muscles, as well as the thymus gland, an organ located behind the breastbone in the upper chest, which helps to regulate the functions and amount of the body's blood cells, including the cells of the immune system. In some cases of myasthenia gravis, cancerous tumors develop in the thymus gland.

In others, the thymus all but disappears. Removal of the gland, in an effort to halt the body's immune attack on itself and the effects of myasthenia on the thymus, is therefore an early step in treating the disease.

AMYOTROPHIC LATERAL SCLEROSIS

Amyotrophic lateral sclerosis, abbreviated as ALS and also known as Lou Gehrig's disease for the well-known New York Yankees baseball player who died of this condition, is a neuromuscular disease that results in the deterioration of motor neurons in the brain and lower spine. As a consequence, there is a gradual loss of control over movement of the affected muscles. The disease is most common in people over the age of 50 and occurs more often in men than in women. Approximately 3,000 new cases of ALS are diagnosed annually in the United States.

Researchers still do not know the source of ALS in most of its victims, although approximately 10% of cases are genetically inherited. Nor, to date, has any cure been developed for the nerve degeneration caused by the disease.

The earliest symptom of ALS may be a weakness in the hands and arms. The disease can also cause muscles to quiver involuntarily and to cramp. Eventually, ALS cripples the neuromuscular function of both arms and both legs and ultimately affects the muscles used for swallowing and respiration, interfering with movement, speech, breathing, and eating. Many patients die within two years after diagnosis of the disease, although some persons with ALS survive for decades.

SPINAL MUSCULAR ATROPHIES

The group of inherited diseases known as spinal muscular atrophies (SMAs) involve the degeneration of nerve cells in the gray matter of the spinal cord. Symptoms vary from one type of SMA to another, but a particularly severe form, Werdnig-Hoffmann paralysis, may be apparent before birth. In this genetically transmitted form of SMA, the fetus exhibits fewer movements than normal, and the infant into which it develops may be weak after birth. In other instances, Werdnig-Hoffmann paralysis first becomes apparent at a later point during infancy. The muscle weakness caused by the disease generally prevents its victims from making all but the slightest leg movements, and children with the disorder commonly lie on their backs in a "frog leg" position. As the disease progresses, the arm and respiratory muscles become involved, with death often occurring within less than a year after the onset of the disease.

CONCLUSION

As science learns more about the musculoskeletal system and reaches a greater understanding of skeletal disorders and nerve-related muscle disease, the human body will have a far better chance of withstanding the biological changes of aging. Scientists working to identify the genetic code behind inherited diseases hope one day to find the cause of SMA, ALS, and other disorders passed from generation to generation. But health-related progress is not limited to the laboratory. If it is possible to increase strength even in individuals who have reached their nineties, then certainly a lifetime of proper diet and exercise can maintain the body's resilient framework well into old age.

APPENDIX: FOR MORE INFORMATION

The following is a list of organizations that can provide information about different disorders that affect the musculoskeletal system.

AMYOTROPHIC LATERAL SCLEROSIS

The ALS Research Center
2351 Clay Street, Suite 416
San Francisco, CA 94115
(415) 923-3604

Amyotrophic Lateral Sclerosis Association
21021 Ventura Boulevard
Suite 321
Woodland Hills, CA 91364
(818) 340-7500

Amyotrophic Lateral Sclerosis Society of Canada
90 Adelaide Street East, Suite B101
Toronto, Ontario M5C 2R4
Canada
(416) 362-0269
(416) 362-0414 (fax)

ARTHRITIS

Arthritis Foundation
1314 Spring Street, NW
Atlanta, GA 30309
(404) 872-7100

Arthritis Society
250 Bloor Street East, Suite 401
Toronto, Ontario M4W 3P2
Canada
(416) 967-1414
(416) 967-7171 (fax)

National Institute of Arthritis and Musculoskeletal and Skin Diseases
9000 Rockville Pike
Building 31, Room 4C05
Bethesda, MD 20892
(301) 496-8188

CANCER

American Cancer Society
19 West 56th Street
New York, NY 10019
(212) 586-8700

Canadian Cancer Society
10 Alcorn Avenue, Suite 200
Toronto, Ontario M4V 3B1
Canada
(416) 961-7223
(416) 961-4189 (fax)

National Cancer Institute
The National Institutes of Health
9000 Rockville Pike, Building 31, 10A18
Bethesda, MD 20892
(301) 496-5583
(800) 4-CANCER

MUSCULAR DYSTROPHY

Muscular Dystrophy Association
810 Seventh Avenue
New York, NY 10019
(212) 586-0808

Muscular Dystrophy Association of Canada
150 Eglinton Avenue East, Suite 400
Toronto, Ontario M4P 1E8
Canada
(416) 488-0030
(416) 488-7523 (fax)

MYASTHENIA GRAVIS

Myasthenia Gravis Foundation
53 West Jackson Boulevard, Suite 660
Chicago, IL 60604
(800) 541-5454

OSTEOPOROSIS

National Osteoporosis Foundation
2100 M Street NW, Suite 602
Washington, DC 20037
(202) 223-2226

SCOLIOSIS

National Scoliosis Foundation
72 Mount Auburn Street
Watertown, MA 02172
(617) 926-0397

Scoliosis Association
(407) 994-4435

SPINA BIFIDA

Spina Bifida Association of America
4590 MacArthur Boulevard, Suite 250
Washington, DC 20007
(202) 944-3285

FURTHER READING

Bastholm, E. *The History of Muscle Physiology: From the Natural Philosophers to Albrecht Von Haller*. Denmark: Ejnar Munksgaard, 1950.

Carola, Robert, John P. Hartley, and Charles R. Noback. *Human Anatomy and Physiology*. New York: McGraw-Hill, 1990.

Dwyer, Anthony P., and Carl L. Nelson, eds. *The Aging Musculoskeletal System: Physiological and Pathological Problems*. Lexington, MA: The Collamore Press, 1984.

Fiatrone, Maria A., et al. "High-Intensity Strength Training in Nonagenarians: Effects on Skeletal Muscles." In *Journal of the American Medical Association*, June 13, 1990.

Guyton, Arthur C. *Textbook of Medical Physiology*. 8th ed. Philadelphia: Saunders, 1991.

Hole, John W., Jr. *Human Anatomy and Physiology*. 5th ed. Dubuque, IA: Brown, 1990.

Jones, D. A., and J. M. Round. *Skeletal Muscle in Health and Disease: A Textbook of Muscle Physiology*. New York: Manchester University Press, 1990.

Kaplan, Paul E., and Ellen D. Tanner. *Musculoskeletal Pain and Disability*. Norwalk, CT: Appleton & Lange, 1989.

Knight, Bernard. *Discovering the Human Body: How Pioneers of Medicine Solved the Mysteries of the Body's Structure and Function*. Philadelphia: Lippincott, 1980.

McArdle, William D., Frank I. Katch, and Victor L. Katch. *Exercise Physiology: Energy, Nutrition, and Human Performance*. 2nd ed. Philadelphia: Lea & Febiger, 1986.

McGrew, Roderick E. *Encyclopedia of Medical History*. New York: McGraw-Hill, 1985.

Szladits, Lola L. "The Influence of Michelangelo on Some Anatomical Illustrations." In *Journal of the History of Medicine*, October 1954.

Tavassoli, Mehdi, and Joseph Mendel Yoffey. *Bone Marrow: Structure and Function*. New York: Alan R. Liss, Inc., 1983.

Wilson, Frank C., ed. *The Musculoskeletal System: Basic Processes and Disorders*. 2nd ed. Philadelphia: Lippincott, 1983.

GLOSSARY

A bands the dark myofilaments located in muscle cells and composed primarily of myosin

acetylcholine a neurotransmitter that carries signals from one nerve cell to another

acetylcholinesterase the enzyme responsible for breaking down acetylcholine after it has finished transmitting its messages

actin a protein found in muscle cells; strands of actin link with myosin to permit muscle contraction

acute beginning suddenly and lasting for a short time

adenosine diphosphate (ADP) a form of adenosine that is reversibly converted to ATP for the storing of energy by the addition of a high-energy phosphate group

adenosine triphosphate (ATP) an enzyme that stores energy needed for muscle contraction in the bonds linking phosphate to the molecule

aerobic living, growing, or occurring only in the presence of oxygen

amyotrophic lateral sclerosis (ALS) a neuromuscular disease that results in the deterioration of motor neurons in the brain and lower spine; also known as Lou Gehrig's disease

anaerobic living, active, or occurring in the absence of free oxygen

analgesic a drug that relieves pain without causing loss of consciousness

anemic a condition characterized by an abnormally low level of red blood cells

arteries tube-shaped vessels that carry blood away from the heart to various parts of the body
arthritis an umbrella term for almost 125 different disorders sharing the common symptom of joint pain and degeneration
atrophy a decrease in size or wasting away of an organ or tissue; muscles will atrophy from lack of use and lose their strength

bipennate muscle a muscle in which the fascicles are arranged in a line along a single bone and fan outward in opposite directions

cardiac muscle striated muscle comprising the heart
cartilage a tough, resilient tissue containing strong fibers of the protein collagen
cellular respiration the process by which muscle fibers turn glucose into fuel for energizing muscle contraction
chromosomes rodlike structures of DNA and protein found in the nuclei of cells; each normal human cell contains 23 pairs of chromosomes
chronic lasting for a long time
citric acid cycle the aerobic process that produces ATP during muscle contraction; the by-products of this cycle are carbon dioxide and hydrogen; also called the Krebs cycle
clubfoot a birth defect, involving muscle and bone, that often gives the foot a clublike appearance
collagen an insoluble fibrous protein that is present in vertebrates as the main component of connective tissue fibrils and bones
compact bone densely structured bone found in the diaphysis of long bones; it forms the covering of all bones in the body
concave hollowed or rounded inward
congenital existing at or dating from birth
convex curved or rounded like the exterior of a sphere
creatine phosphate molecule that serves as an energy source for the production of ATP during muscle contraction
cross union the fusion of two bones as they heal after a fracture

diaphragm a wall of muscle, located in the upper abdomen, whose regular contractions are responsible for the expansion of the lungs and the inhalation of air, and whose relaxations compress the lungs and force them to exhale the air they have inhaled
diaphysis the shaft of a long bone
dislocation the failure of the ends of the bones in a joint to make proper contact with each other

Duchenne muscular dystrophy a severe form of muscular dystrophy inherited as a sex-linked recessive trait; it affects the shoulders and the pelvic girdle

endoskeleton an internal skeleton or supporting framework

enzyme any of numerous complex proteins that are produced by living cells and catalyze specific biochemical reactions at body temperature

epiphysis the section of bone at the end of a long bone

erythrocytes red blood cells

estrogen a sex hormone produced in the adrenal glands of both sexes, in the ovaries of women, and in placentas; regulates reproductive function and secondary sex characteristics

exoskeleton an external supportive and protective covering of an animal

external fixation a means of immobilizing a fracture in which the bone is held in place by attaching a rigid supporting device to the outside of the body at the site of the break

fascia a sheet of connective tissue that protects the muscle and contains nerves, blood vessels, and other structures that serve the muscle

fascicle a small bundle of muscle fibers

fast-twitch fibers type of fibers in the skeletal muscle that are best for perfoming rapid muscle contractions over a short period of time; they tend to tire quickly

fibers the threadlike structures that compose the muscles

flat bones broad, thin bones, such as the ribs

genes complex units of chemical material contained within the chromosomes; variations in the patterns formed by the components of genes are responsible for inherited traits

glycolysis the first phase of cellular respiration, in which one molecule of glucose, released from glycogen, is broken down to form two molecules of pyruvic acid and four molecules of ATP; this process is anaerobic

hemoglobin a pigment found in red blood cells that transports oxygen from the lungs to the body's tissues

hyaline cartilage smooth, bluish white connective tissue used to cushion joints

hypertrophy an increase in muscle size resulting from an increase in the diameter of each muscle cell

I bands the light myofilaments in muscle cells, composed primarily of actin

idiopathic arising from an obscure or unknown source
insertion the point of a skeletal muscle's attachment to the bone that the muscle moves
internal fixation immobilization of a fracture in which the fractured bone is held in place by a metal plate or pin attached directly to the fractured bone
irregular bones bones with a number of different shapes
isometric contraction a contraction that occurs when a muscle tightens but the body part is unable to move; often happens during an attempt to lift a heavy object that will not budge
isotonic contraction a contraction that enables a body part to move and, in turn, move an object

lactic acid a substance formed when glucose in the muscles is broken down anaerobically
leukocytes white blood cells
ligaments fibrous bands of tissue that connect one bone to another in the region of a joint
long bones bones that have a diaphysis, making them longer than they are wide
long muscles the muscles found in the back that are necessary to maintain posture; they consist primarily of slow-twitch fibers

malunion when a fracture heals in an abnormal position or in poor alignment
marrow soft tissue found in spongy bone and the medullary cavity of long bones
matrix in the skeleton, the material found between bone cells, where calcium salts and other compounds are stored
medullary cavity canal running longitudinally through a long bone; contains yellow marrow
menopause the natural cessation of menstruation, usually occurring between the ages of 45 and 50
metabolism the process by which substances within a living organism are chemically broken down in order to release energy
mitochondria structures located within muscle fibers that manufacture ATP, which energizes muscle contraction
motor end plate the flat region at the end of a motor neuron fiber where the fiber meets the surface of a muscle cell
motor neuron a nerve cell that transmits impulses for muscle contraction
motor unit a motor neuron and the muscle cells to which its fibers connect
multipennate muscle a muscle in which the fascicles fan out from a common point

muscle fiber muscle cell
muscular dystrophy a disease that causes a progressive wasting of the muscles
musculoskeletal system the network of muscles and bones used for support, movement, and metabolic functions of the body
myasthenia gravis an autoimmune disease caused by a defect in the neuromuscular junction; characterized by a progressive weakness of the voluntary muscles without atrophy or sensory disturbance
myeloma the most common type of bone cancer, in which malignant tumors in the bone marrow hinder the generation of red blood cells
myofibrils long, thin structures in muscle fibers made from folds in the membranes of the sarcoplasmic reticulum
myofilaments one of the individual filaments of actin or myosin that make up a myofibril
myoglobin a pigment in muscle cells that is used to store oxygen
myopathy any disorder of muscle tissue
myosin a protein found in muscle cells; strands of myosin link with actin to permit muscle contraction

neuromuscular junction the intersection of a nerve and muscle fiber
neurotransmitter a chemical that carries nerve signals across gaps between nerve cells or between nerve cells and muscle cells
nonunion failure of the fractured ends of a bone to grow back together

origin the site at which a skeletal muscle is attached to a bone that does not move
orthopedics the branch of medicine specializing in the prevention and treatment of skeletal disorders, muscular ailments, and related problems
osteoarthritis the most common form of arthritis; usually a result of repeated stress on the body's joints over the course of a lifetime
osteoblast a cell used in the formation of the bone matrix
osteoclast large cells with multiple nuclei, found in bone and responsible for removing excess bone tissue
osteocyte a mature bone cell
osteogenic cell a bone cell that can be transformed into either an osteoblast or an osteoclast
osteogenic sarcoma a form of bone cancer that occurs in adolescents during the years when they undergo their greatest growth spurts
osteomalacia a disease that affects adults, characterized by soft and deformed bones and due to a lack of vitamin D
osteoporosis a disease characterized by a decrease in the formation of new bone while the body continues to resorb old bone at its normal rate; it leaves bones porous and easily breakable

oxygen debt following muscle activity, the amount of ATP needed to convert all the lactic acid in the liver to glucose and to return ATP and creatine phosphate levels in the body to their pre-exercise levels

Paget's disease a disease in which bone is both destroyed and replaced at an excessive rate, leading to the enlargement of the affected portion of the skeleton, with weakening of the diseased bone
periosteum the fibrous membrane that surrounds the bones
peristalsis the process in which smooth muscle lining the stomach and the intestines contracts to push food through the digestive system
pituitary dwarf a person who fails to produce enough human growth hormone; the body is normally proportioned but extremely small in size
pituitary giant a person who secretes an excessive amount of human growth hormone, causing the person's height to be greater than normal
platelets blood cells that control clotting; also called thrombocytes
Pott's disease tuberculosis of the spine, caused by the bacteria *Myobacterium tuberculosis*

recessive a characteristic that will appear in an offspring only if the child received the genes of that trait from both parents
red marrow a form of marrow that produces oxygen-carrying red blood cells as well as some of the immune system's white blood cells; found in spongy bone
red muscles muscles characterized by a strong blood supply and a large quantity of myoglobin; usually contain slow-twitch fibers
reduction the realignment of broken bones so they heal in the correct position
resorb to break down and assimilate something that was previously differentiated
rheumatoid arthritis a chronic disease characterized by pain, stiffness, inflammation, swelling, and sometimes destruction of joints
rhythmic contractions pulsed contractions that travel through smooth muscle to keep food moving along the digestive tract
rickets a disease characterized by soft and deformed bones and caused by a vitamin D deficiency that affects the young during the period of skeletal growth
round bones small bones that are usually found within tendons

sarcolemma the membrane surrounding a skeletal muscle fiber
sarcoplasm semifluid material found inside a muscle fiber
sarcoplasmic reticulum the series of membranes, small fluid-filled sacs, and small canals within the sarcoplasm of muscle cells
scoliosis a side-to-side curvature of the spine that affects infants, children, and adolescents and becomes progressively more severe

sesamoid bones bones commonly found within tendons adjacent to joints
short bones bones that are roughly as long as they are wide
skeletal muscle a muscle that is attached to bone and is responsible for the body's voluntary movements; skeletal muscle fibers are striated
slow-twitch fibers fibers in the skeletal muscle that generate more ATP than fast-twitch fibers because they work at a slower rate; best suited for aerobic respiration
smooth muscle a muscle responsible for many automatic movements, including contraction of the intestines during digestion; unlike skeletal muscle cells, smooth muscle fibers are not striated
spina bifida a spinal deformity resulting from the incomplete closure of portions of the vertebrae around the spinal cord
spongy bone bone structured to include beamlike processes for support containing numerous marrow-filled spaces
sprain an injury in which the ligaments connecting two bones at a joint are stretched or torn when the joint is twisted.
striations dark and light bands found in skeletal muscle fibers formed from strands of actin and myosin
suture the immovable fibrous joint where the flat bones of the skull meet

tendon a fibrous band of connective tissue attaching muscle to bone
testosterone the male hormone responsible for inducing and maintaining male secondary sex characteristics, such as a deep voice and facial hair
thrombocytes blood cells that control clotting; also called platelets
tonic contractions contractions that keep the muscle at least partially contracted at all times

unipennate muscle a muscle in which the fascicles begin as a close group on one bone and then angle outward to connect along the length of another bone

veins vessels that carry blood from various parts of the body back to the heart
vertebrae bones that compose the backbone—there are 33 in an infant and 26 in an adult—and are bound together by ligaments and intervertebral disks
vertebrates animals that have backbones

white muscles muscles characterized by a smaller blood supply and a more limited quantity of myoglobin; they usually contain fast-twitch fibers

yellow marrow a form of bone marrow that functions as a storage area for fat; found in the medullary cavities

INDEX

A bands, 46
Acetylcholine, 47, 48, 93
Acetylcholine receptors, 92
Acetylcholinesterase, 48, 93
Acetyl-coenzyme A, 50
Actin, 46, 47, 48, 49, 56, 61, 62
Adenosine diphosphate (ADP), 49
Adenosine triphosphate (ATP), 49, 50, 51, 52, 53, 58, 62
Aerobic reaction, 50
Albee, Fred H., 20–21
Alcohol, 74, 76
Alpha-fetoprotein, 81
Amyotrophic lateral sclerosis, 94
Anaerobic reaction, 50
Analgesic, 78
André, Nicholas, 19–20
Anemia, 27, 82
Antibodies, 92
Aponeuroses, 46
Arteries, 14, 56
Arterioles, 56
Arthritis, 33, 76–81
Aspirin, 78, 80
Atlantoaxial joint, 31
Atrophy, 39, 62
Auditory nerve, 84
Autoimmune disease, 79

Baglivi, Giorgio, 20
Ball-and-socket joint, 31–32
Becerra, Gasparo, 16–17
Becker muscular dystrophy, 90
Biceps muscle, 55, 60
Bipennate muscle, 56
Blood pressure, 56
Bone cancer, 82
Bone fracture, 65–71
Bone graft, 21
Bone *matrix*, 26, 29, 36, 38, 41
Bones
  aging and, 41–42
  blood cell production and, 27
  development, 35–37
  disease and,74–86
  early treatment advances, 19–21
  function, 24–27
  growth, 38–41
  historical study of, 14–19
  injuries, 65–71
  structure, 29–30
  types, 27–28
Bonesetter, 17

Calcitonin, 40, 47, 76, 85
Calcium, 25, 26, 30, 38, 40, 41, 42, 48, 54, 74, 83, 90
  disease and, 41, 76
  function, 25
Calcium carbonate, 25
Calcium phosphate, 25
Calmodulin, 57

Canaliculi, 29, 30
Capillaries, 63
Cardiac muscle fiber, 58
Carpal bone, 27
Cartilage, 29
Cartilaginous plate, 38
Casts, 69–70
Cellular respiration, 50–51
Central nervous system, 16, 61
Chromosomes, 89–90
Citric acid, 50
Citric acid cycle (Krebs cycle), 50, 51
Clavicle, 68
Clubfoot, 20
Coenzyme A, 50
Collagen, 29, 38, 41
Comminuted bone fracture, 66
Compact bone, 29, 36, 38, 42, 68
Compound bone fracture, 66
Condyloid joint, 33
Congenital defect, 72, 86
Connective tissue, 68
Cortical bone, 42
Corticosteroids, 78, 80
Creatine phosphate, 49–50, 52
Cross union, 69
Cytochromes, 51
Cytosine, 91

Da Vinci, Leonardo, 15–16
De Humani Corporis Fabrica (The Fabric of the Human Body)(Vesalius), 17, 19
Depressed bone fracture, 66
Diaphragm, 43
Diaphysis, 29, 36, 37, 42
Digestive tract, 43, 56, 58
Duchenne muscular dystrophy, 89–91
Dystrophin, 90

Electron transport system, 51
Endochondral bone, 36
Endochondral ossification, 36–37
Endomysium, 45, 46
Endoskeleton, 24
Endosteum, 68
Enzyme, 48
Epimysium, 46
Epiphyseal line, 38
Epiphyseal plate, 34, 37, 38, 40, 83
Epiphysis, 29, 30, 37
Erythrocytes, 27
Estrogen, 42, 74, 76
Exercise
    elderly and, 64
    muscle development and, 63–64
    muscle endurance and, 62
    muscle fatigue and, 52
    muscle size and, 61
    muscle strength and, 61–62
    osteoporosis and, 75
    oxygen debt and, 51–52
Exoskeleton, 24
External fixation, 69

Facioscapulohumeral dystrophy, 91
Fascia, 46
Fascicles, 45–46, 55–56
Fast-twitch fiber, 52–53, 63
Femur, 75
Fibers, 45, 46, 52–53
Fibroblasts, 68
Fibrocartilage, 34
Fibrocartilaginous callus, 68
Fibula, 27, 34
Fissured bone fracture, 66
Flat bone, 28, 30
Fusiform muscle, 55

Galen, 14–15, 18, 19
Gastrointestinal tract, 40
Gliding joint, 31
Glucose, 50, 52
Glycogen, 50, 53
Glycolysis, 50, 52
Gomphosis joint, 34
Granulation tissue, 68
Greenstick bone fracture, 66
Growth hormone, 39–40, 54

Guanine, 91
Guyton, Arthur C., 61

Haversian canals, 29, 30
Heart, 14, 16, 24, 58
Hemoglobin, 27
Hinge joint, 33
Hippocrates, 15
Hormones, 39–40
Hyaline cartilage, 31, 34, 36
Hydroxyapatite, 26
Hyoid bone, 55
Hypertrophy, 61

I bands, 46
Idiopathic disease, 86
Immune system, 79–80, 92
Impacted bone fracture, 66
Intercalated disks, 58
Intercellular material, 36
Internal fixation, 69, 71
Interosseous ligament, 33
Intervertebral disk, 34
Intramedullary nail, 71
Intramembranous ossification, 35–36
Irala, Mateo Antonio, 17
Irregular bones, 28, 30
Isometric contractions, 61
Isotonic contractions, 61

Joint, 14, 24, 29, 31–34
  cartilaginous, 34
  disease and, 76–81
  dislocation, 71–72
  fibrous, 33–34
  function, 29
  synovial, 31–33
Joint capsule, 31

Krebs, Hans Adolph, 50
Kunkel, Louis, 90
Kyphosis, 75, 86

Lactic acid, 21, 51, 52
Lacunae, 29, 30, 36
Lamella, 29
Leeuwenhoek, Antonie van, 20
Leukocytes, 27
Ligaments, 29, 31, 72
Liver, 51–52
Long bones, 27, 83
  formation, 36
  structure, 29
Long muscles, 53
Lungs, 16, 24, 43

Malunion, 69
Marrow, 15, 20
Maternal serum alpha-fetoprotein screening test, 81
Medullary cavity, 27, 29, 36, 38, 71
Menopause, 42, 74
Metabolism, 24
Metacarpal bones, 33
Methotrexate, 80
Mitochondria, 49, 53, 58, 62, 63
Motor end plate, 47
Motor neurons, 46–47, 94
Motor unit, 47, 88
Multipennate muscle, 56
Muscle, 43–58
  aging, 53–54
  cardiac, 58
  disorders, 87–95
  exercise and, 51–52, 59–64
  function, 43, 54–55
  historical study of, 14–21
  shape, 55–56
  skeletal, 45–47
  smooth, 56–58
Muscle contraction, 24, 25, 47–49, 60–61
Muscle cramps, 88
Muscle impulse, 47
Muscle motor units, 54
Muscle tone, 59
Muscular dystrophy, 21, 88–91
Myasthenia gravis, 92–94
Myeloma, 82

Myofibrils, 46
Myoglobin, 51, 53, 63
Myopathies, 89
Myosin, 46, 47, 48, 49, 54, 56, 61, 62
Myosin adenosine triphosphatase, 49
Myotonic dystrophy, 91

Neumann, Ernst, 20
Neural tube defects, 81
Neuromuscular junction, 47
Neurotransmitters, 47
Nonunion, 69

Oblique bone fracture, 66
Orbicularis oculi muscle, 54
Orbicularis orbis muscle, 54
Orthopedics, 19–21
Osteoarthritis, 76–78
Osteoblasts, 36, 38, 68
Osteoclasts, 26, 36, 38, 40, 68
Osteocytes, 30, 36
Osteogenic sarcoma, 82
Osteomalacia, 41, 83–84
Osteomyelitis, 83
Osteoporosis, 74–76, 82
Ovaries, 39
Oxaloacetic acid, 50
Oxygen, 27, 50, 51, 58
Oxygen debt, 52

Pacemaker cell, 57
Paget, James, 84
Paget's disease, 84–85
Pannus, 79
Parathormone, 40
Parathyroid gland, 39, 40, 41, 42, 74
Parathyroid hormone, 74, 76
Patella, 28
Pelvic girdle, 28
Perichondrium, 36
Perimysium, 45, 46
Periodontal ligament, 34
Periosteum, 29, 36, 45, 68
Peristalsis, 56
Phagocytes, 68
Phosphate, 41, 49
Phosphorus, 41, 74, 76
Piezoelectricity, 71
Pituitary dwarf, 40
Pituitary giant, 40
Pituitary gland, 39, 40
Pivot joint, 31
Plaster of paris, 20, 69–70
Platelets, 27
Polyurethane, 70
Pott, Percivall, 20
Pott's disease, 20
Primary ossification center, 36
Proximal phalanges, 33
Pseudoarthrosis, 69
Pyruvic acid, 50, 51

Red marrow, 27, 30, 37
Red muscle, 53
Resorption, 41
Rheumatoid arthritis, 72, 79–81
  cause, 79–80
  treatment, 80–81
Rhythmic contractions, 58
Ribs, 24, 28
Rickets, 41
Round bones, 28
Rufus of Ephesus, 14

Saddle joint, 33
Sarcolemma, 46, 47
Sarcolemmal membrane, 47, 92
Sarcoplasm, 46, 47
Sarcoplasmic reticulum, 46, 47, 48, 49, 54, 56
Scoliosis, 85–86
Shin splint, 88
Short bones, 27, 30
Shoulder girdle, 24
Simple bone fracture, 68
Skeletal muscle, 45–47
Skull, 24, 28
Sliding-filament theory, 48

Slow-twitch fiber, 52–53, 63
Smooth muscle, 20, 56–58
Sphincter muscle, 54
Spina bifida, 81–82
Spinal column, 28, 74–75, 81, 85–86, 94
Spinal muscular atrophies, 94–95
Spiral bone fracture, 66
Splints, 70
Spongy bone, 30, 36, 37, 42, 68
Sprains, 72
Staphylococcus aureus, 83
Stefan, Jan, 17
Sternohyoid muscle, 55
Sternum, 28
Stomach, 56, 58
Subluxation, 72
Suture, 33
Sylvius, Jacobus, 18, 19
Symphysis joint, 34
Synchondrosis joint, 34
Syncytium, 58
Syndesmosis, 33–34
Synovial fluid, 72
Synovial membrane, 72

Tarsal bones, 27
Temporomandibular joint, 72
Tendons, 14, 28, 45, 46
Testes, 40
Testosterone, 61
Thomas, Hugh Owen, 20
Thomas splint, 20
Thorax, 75
Thrombocytes. *See* Platelets
Thymine, 91
Thymus gland, 93–94
Thyroid gland, 39, 40, 76
Tibia, 27, 34
Tonic contractions, 58
Trabeculae, 30
Trabecular bone, 42
Transverse bone fracture, 66
Trapezius muscle, 56
Tropomyosin, 48–49
Troponin, 48–49

Ulna, 27
Unipennate muscle, 56
Urinary bladder, 58

Valproic acid, 81
Vertebrae, 24, 28, 37, 81, 84, 86
Vesalius, Andreas, 17–19
Vitamin A, 41
Vitamin C, 41
Vitamin D, 40, 83, 84
Von Helmholtz, Hermann, 21

Werdnig-Hoffman paralysis, 94–95
White muscles, 53

Yellow marrow, 26–27, 37

PICTURE CREDITS

American College of Rheumatology: p. 77; Courtesy Department of Library Services, American Museum of Natural History, neg. #325985: p. 35; The Arthritis Foundation: pp. 73, 78; The Bettmann Archive: pp. 15, 44; Library of Congress: pp. 13, 22; Muscular Dystrophy Association: pp. 87, 89; NASA: p. 39; National Library of Medicine: pp. 16, 18, 21, 51, 85; Photo Courtesy Pfizer, Inc.: p. 80; Reuters/Bettmann: pp. 59, 63; United Nations: p. 75; University of California, Lawrence Berkeley Laboratory: p. 92; UPI/Bettmann: pp. 40, 52, 65; Original illustrations by Wendy Beth Jackelow: pp. 23, 25, 26, 28, 30, 32, 37, 43, 45, 55, 57, 60, 62, 67, 70

Brian Feinberg is a Connecticut–based editor and author. Prior to entering book publishing he spent more than five years as a reporter for the *Commercial Appeal*, a daily newspaper in Memphis, Tennessee, and worked as an associate health editor for *Redbook* magazine. Feinberg is currently a life sciences editor for the *Encyclopedia Americana.*

Dale C. Garell, M.D., is medical director of California Children Services, Department of Health Services, County of Los Angeles. He is also associate dean for curriculum at the University of Southern California School of Medicine and clinical professor in the Department of Pediatrics & Family Medicine at the University of Southern California School of Medicine. From 1963 to 1974, he was medical director of the Division of Adolescent Medicine at Children's Hospital in Los Angeles. Dr. Garell has served as president of the Society for Adolescent Medicine, chairman of the youth committee of the American Academy of Pediatrics, and as a forum member of the White House Conference on Children (1970) and White House Conference on Youth (1971). He has also been a member of the editorial board of the *American Journal of Diseases of Children.*

C. Everett Koop, M.D., Sc.D., is former Surgeon General, deputy assistant secretary for health, and director of the Office of International Health of the U.S. Public Health Service. A pediatric surgeon with an international reputation, he was previously surgeon-in-chief of Children's Hospital of Philadelphia and professor of pediatric surgery and pediatrics at the University of Pennsylvania. Dr. Koop is the author of more than 175 articles and books on the practice of medicine. He has served as surgery editor of the *Journal of Clinical Pediatrics* and editor-in-chief of the *Journal of Pediatric Surgery*. Dr. Koop has received nine honorary degrees and numerous other awards, including the Denis Brown Gold Medal of the British Association of Paediatric Surgeons, the William E. Ladd Gold Medal of the American Academy of Pediatrics, and the Copernicus Medal of the Surgical Society of Poland. He is a chevalier of the French Legion of Honor and a member of the Royal College of Surgeons, London.